The AltaVista Search Revolution, Second Edition

The AltaVista Search Revolution, Second Edition

Eric J. Ray,
Deborah S. Ray,
and Richard Seltzer

Osborne **McGraw-Hill**

Berkeley New York St. Louis San Francisco
Auckland Bogotá Hamburg London Madrid
Mexico City Milan Montreal New Delhi Panama City
Paris São Paulo Singapore Sydney
Tokyo Toronto

Osborne/McGraw-Hill
2600 Tenth Street
Berkeley, California 94710
U.S.A.

For information on translations or book distributors outside the U.S.A., or to arrange bulk purchase discounts for sales promotions, premiums, or fund-raisers, please contact Osborne/**McGraw-Hill** at the above address.

The AltaVista Search Revolution, Second Edition

1234567890 AGM AGM 901987654321098

ISBN 0-07-882435-4

Publisher
Brandon A. Nordin

Editor-in-Chief
Scott Rogers

Acquisitions Editor
Megg Bonar

Project Editor
Heidi Poulin

Editorial Assistant
Stephane Thomas

Technical Editor
Danny Sullivan

Copy Editor
Vivian Jacquette

Proofreader
Stefany Otis

Indexer
Rebecca Plunkett

Computer Designer
Jani Beckwith
Michelle Galicia
Jean Butterfield
Marcela Hancik
Roberta Steele

Illustrator
Brian Wells

Series Design
Michelle Galicia

About the Authors . . .

Eric J. Ray and Deborah S. Ray, co-authors of award-winning titles such as *Mastering HTML 4.0, HTML 4 for Dummies Quick Reference, Dummies 101: HTML 4, Netscape Composer for Dummies*, own RayComm, Inc., a technical communications consulting firm.

Eric has been involved with the Internet for over five years and has made numerous presentations and published several papers about HTML and online information. His technical experience includes creating and maintaining the TECHWR-L listserv list as well as implementing and running Web servers.

Deborah, a technical communicator for the past four years, has been involved with the Internet for the past three. Her technical experience includes writing various computer and engineering documents, including the documentation package for a radar signal switchboard used by the U.S. Navy. Deborah currently teaches technical writing at Utah State University and previously taught the same at Oklahoma State University.

Eric and Deborah can be reached at **ejray@raycomm.com** and **debray@raycomm.com**, or through their Web site at **http://www. raycomm.com/**.

Richard Seltzer is a marketing consultant in the Internet Business Group at Digital. He frequently speaks on Internet topics, acting as an advocate for full and more effective use of the Internet for business and education. In addition, Richard runs his own small publishing business on the Internet (The B&R Samizdat Express), and is the author of novels (*The Name of Hero*) and children's books (*The Lizard of Oz*). His electronic newsletter, Internet-on-a-Disk, has a readership of over 100,000, and his acclaimed Web site (**http://www.samizdat.com/**) is frequently cited as an important resource for education and the blind. Richard also writes a monthly newsletter, the *Internet Search Advantage*, which can be found at **http:www.cobb.com/isa**. His next Internet book is entitled *The Social Web: From Hyperlinks to People Links*. A graduate of Yale ('69), he has a master's degree in comparative literature (Russian, French, and German) from the University of Massachusets. Richard lives in Boston with his wife Barbara and four extraordinary children. Contact Richard with any questions or comments at **seltzer@samizdat.com** or **richard.seltzer@digital.com**.

CONTENTS AT A GLANCE

CONTENTS

FOREWORD

On December 14, 1995, one day before the formal introduction of AltaVista to the Internet community, a few of us in Palo Alto watched in awe as the word about AltaVista's existence began to spread like wildfire and the traffic at the site climbed ever higher. I knew then that we were making history and that the need for information was clearly much larger than what had been previously available.

Within weeks, hundreds of articles were written on how AltaVista's Super-Spider could help you find long-lost friends or list all sites pointing to your page, and the public could not get enough of it.

With a small group of friends and colleagues, we had done in a few short months what was then established as clearly impossible: indexing the entire Web and serving it to millions of users in a fraction of a second. The ingredients for success: incredibly talented colleagues, the creative talent, the freedom and infrastructure of the research laboratories of Digital, a deep understanding of the Internet, much midnight oil, a very patient wife, and perfect timing.

By the end of January 1996, it was clear that AltaVista was here to stay. It wasn't just another Web site or another Web-based service—it had radically changed the nature of the Internet and how people use it. A group of Digital managers visualized a book about AltaVista, and here began my interaction with the authors of this book.

I've greatly enjoyed working with these authors on both the original version of the AltaVista Search Revolution and this new Second Edition. Richard Seltzer is a familiar face from within Digital—it was from his original outline that this book was built. Richard interviewed me and the entire AltaVista development team to get a sense of how we built AltaVista into what it is today. Richard also contributed some early research results on how AltaVista was being used and talked about in newsgroups and contributed valuable insight for what readers might like to know.

Eric and Deborah Ray have done a great job of showing you how to use the various parts of AltaVista. Their tips on how to make your searches find exactly what you want, along with ideas for just what you can search for combine to make this book easy to read and to understand. Eric and Deborah have added the fine-tuning on the inside story of AltaVista as the project has evolved; this edition's Chapter 10 provides both a good look at how we've arrived at the present and a unique look at a project whose implications none of us could

predict when we began back in 1995. In this way, the book provides one more step in AltaVista's ongoing process of making information on the Internet as widely accessible as possible.

Almost three years after the site went live, AltaVista is *still* making history. It is the most used search engine on the planet. Though the site now offers more content and functions, it is still very much centered around Web searches. The Web has grown to maybe a billion documents, depending how one counts. It has, however, not become any more organized. The old saying "the Web is only as good as its index" is just as true now as it was when AltaVista was introduced. Through constant investment in technology, AltaVista is continuing to offer access to the best index of the Web.

Louis Monier
Senior Consulting Engineer
Digital Equipment Corporation

ACKNOWLEDGMENTS

Creating this book was a collaborative effort and involved the help of many people.

We'd like to thank several people at Osborne/McGraw-Hill for their efforts in making this book as good as possible. A big thank you goes to Megg Bonar and Scott Rogers for helping develop, refine, and manage the overall project. This book would not have been possible without their help and contributions. Also, thanks to Stephane Thomas and Heidi Poulin for bringing order to our manuscript.

We'd also like to thank the entire AltaVista team and the folks at Digital's research labs for providing information and context for this book. In particular, thanks to Kathleen Greenler for providing feedback and direction and to Louis Monier for providing technical information for this book. Also, thanks to Stephen Stuart, Paul Flaherty, and Phil Steffora for providing additional information and insight into how AltaVista works and how it came about.

Finally, thanks to the thousands of people who provided examples of AltaVista use.

INTRODUCTION

Welcome to *The AltaVista Search Revolution, Second Edition*! Just as the World Wide Web revolutionized the Internet by making a wealth of information easily available to anyone with Web browser software, Digital Equipment Corporation's AltaVista Search service continues to revolutionize the Web by making information easily available *and* easy to find.

For example, suppose you have a few items left in your refrigerator and want recipes for interesting dishes you could make. With a cookbook, you would first have to guess the types or the names of the dishes, then try to find recipes. With AltaVista, you simply enter the items that you have on hand and perhaps the word "recipe" as well.

Or, perhaps you are plagued by a few words of what may be a song, a poem, or a passage from a book. If you know the author, that it is a "classic," and the first words of the famous quote, you'd use Bartlett's or another reference work. But, often, you don't know who wrote it or when, and these are random words from the middle of something. With AltaVista, you just enter the words that you know.

Maybe your computer just crashed and a cryptic error message appeared on the screen as it died. You don't know if it's a hardware or a software problem, nor can you even guess what the problem could be. You don't know what documentation to use and where to look in it. But, with AltaVista, you can just enter the error message and let AltaVista do the searching for you.

And that's not all. AltaVista revolutionizes Internet searching by giving you fast, precise results, but also provides a slew of extra services, such as People and Business Search, Search by Category, and language translation features. These features, combined with AltaVista's scope, speed, and ease of use make AltaVista the most versatile, most user-friendly Internet service available.

The purpose of this book is to help you take full advantage of AltaVista's services and how to operate in the new and changing Internet environment, as both an information consumer and an information provider. You can use AltaVista as a tool to find long-lost friends or relatives and meet new friends; to check on what your competitors are doing, what your customers are saying, and to find new markets; to complete tonight's homework assignment or do basic research for your doctoral dissertation.

The AltaVista Search Revolution, Second Edition will give you a guided tour of AltaVista from its origins in Digital Equipment Corporation's Palo Alto,

California, laboratories to its current place as the preeminent Internet search service serving over 32 million searches each and every day.

Who Should Use This Book

You should read this book if you want to use the Internet to find anything you want, anytime you want. If you can't find the information you need using AltaVista, you won't find it out there. Other search and directory services provide useful links to information on the Web or Usenet newsgroups, but AltaVista is the service that indexes more of the Web and Usenet than any other search service and indexes *every* word it finds out there. AltaVista also contains the largest and most comprehensive index of current newsgroups, not to mention features, such as People and Business Search, Search by Category, and translation services, that make AltaVista a complete Internet search tool. With over 100 million indexed pages and over 32 million queries served each day, no other Internet search tool can match it.

What to Remember About This Book

AltaVista continues to change and evolve—sometimes on a daily basis. For that reason, you may find that some of the figures and content described in this book do not precisely match AltaVista's latest developments. However, the cosmetic changes notwithstanding, you will find that the basic instructions for finding and providing information remain the same. Additionally, of course, the history of AltaVista and the examples of AltaVista's scope and breadth also hold true, even though additional features or services may appear at the AltaVista site.

What's In This Book

This book contains the following types of information:

- An introduction to AltaVista and Internet Searches.
- Step-by-step instructions and useful descriptions of AltaVista features.
- Tips for Web masters and information providers on how they can take advantage of the power of AltaVista.
- The AltaVista story, including inside information about AltaVista's development.
- The AltaVista A to Z Reference, complete with examples and anecdotes.

Chapter One introduces AltaVista. We describe how Internet search engines and directories work and explain their role in the development of the Internet. You'll see why Alta Vista is the premier search engine on the Internet.

Chapter Two shows you how to get great results using AltaVista Simple Search. You'll meet the AltaVista home page, get started with Simple Searches, and be on your way to finding any piece of information on the Internet.

Chapter Three shows you how to use the People and Business Search features—a must for anyone looking up people or businesses on the Internet. You'll see that, with just a wee bit of your input, AltaVista can find the information you're looking for.

Chapter Four explains how to use AltaVista's Refine feature, which makes narrowing your searches a cinch. You'll see how to refine your searches and find the information you need with just a click of your mouse.

Chapter Five introduces AltaVista Advanced Search, which offers more power and options than Simple Search. You'll see that entering operators and using grouping techniques is easy and produces precise results.

Chapter Six presents both Simple and Advanced Searches of Usenet newsgroups. In addition to explaining how newsgroups differ from the World Wide Web, you'll see how to search the 20,000+ newsgroups with their millions of messages.

Chapter Eight presents information providers—whether you're a Web master at a large corporation or an individual with a personal Web site—with the tips and techniques necessary to ensure that AltaVista indexes your pages effectively. You'll also find useful information and tips to help your readers find and use your Web pages effectively.

Chapter Nine shows you how easy AltaVista makes Internet searching by presenting dozens of subject categories, a brief description of the category, followed by a list of specific search ideas and samples.

Chapter Ten presents the story of how AltaVista came to be—how it was possible to create a single tool that can do it all and how a large company was able to turn an R&D project into such a powerful and useful tool for Internet users everywhere.

Appendix A shows a few seconds of incoming AltaVista searches.

Appendix B provides lists of commonly searched for words.

Appendix C provides a list of the most common words published on the Web.

Appendix D shows behind-the-scenes photos of AltaVista.

What You'll Need To Use AltaVista

Before you can use AltaVista, you'll need to have a computer with an Internet connection and some kind of Web browser software. That's it!

Icons Used In This Book

Throughout this book, you'll see a few icons that indicate special information.

Tip *You'll see this icon in places where we provide extra information making AltaVista a little easier to use. Most of this information we've found out through trial and error and want to pass it on to you.*

Note *Where you see this icon, you'll find background or supplementary information. This information is not absolutely essential for you to use AltaVista, but it will help you better understand AltaVista as a whole.*

Remember *Information following this icon will remind you of concepts or steps you learned in previous sections or chapters. We provide these little reminders so that you won't have to keep flipping back to other chapters for information.*

Enjoy *The AltaVista Search Revolution!*

Introduction to AltaVista

Searching the Internet will never be the same! The AltaVista revolution continues to transform the Internet into a medium that you can use to find the information you want when you want it. Even with over 100 million documents at your fingertips, you'll be able to find just the item you're looking for. Gone are the days of arcane UNIX-based commands, the repeated trips to different directories and search engines, the hours of poring over irrelevant results, and the frustration of knowing that the data you need is probably out there, somewhere—if only you could *find* it. With a single search using AltaVista, you can search through the World Wide Web or through over 20,000 Usenet newsgroups to find that piece of information you need.

What makes Digital Equipment Corporation's AltaVista so special? Scope, speed, ease of use, and innovation. AltaVista indexes the World Wide Web and Usenet newsgroups and makes that index easily available to the entire Internet community. It provides a single entry point and simple interface—just type a few words and click Search—to the World Wide Web and 20,000-plus Usenet newsgroups and gives you quick access to all the information they contain. AltaVista empowers you by giving you, in seconds, information that might otherwise be impossible to find.

Since AltaVista debuted, instead of wading through pages and pages of information from hundreds or thousands of sources, you now only have to search one location to quickly find the information you need. AltaVista *is* the standard to which all other Internet search tools are compared. If you can't find it using AltaVista, it's probably not out there.

During the past two years of AltaVista's development, many other search services have adapted AltaVista features and approaches to Internet searching problems. AltaVista, however, has continued to innovate and grow, adding features like Refine (to help focus your search) and a language translation service, making AltaVista's technology cutting edge—and revolutionary.

This chapter explains why AltaVista is so exciting and provides some background to make it easier to see the full significance of the AltaVista revolution. First, you will learn how AltaVista differs from other search tools and how it continues to transform the Internet and business in general. Next, you will see how the Internet developed, the immense problems with Internet searches, and other Internet search solutions. Finally, you will learn how AltaVista really does stand alone as *the* Internet search tool.

This knowledge will prepare you for the rest of this book, which describes some of the tools and techniques used to find order in chaos, and you'll see how you can use AltaVista to both access and provide access to the information available through the Internet.

ALTAVISTA: THE REVOLUTION!

It's quick . . . it's easy . . . it's convenient . . . it's here . . . *and it keeps getting better.*

AltaVista solves the problems of finding information on the Internet and brings the information right to your desktop. AltaVista is *not* just another search service; rather, its scope, speed, ease of use, and innovation make it *the* search engine of the Internet.

AltaVista's Scope

AltaVista, developed in 1995 by Digital Equipment Corporation in its Palo Alto labs, did the unthinkable—it indexed the Internet in a project of unmatched scope. As other search service providers also developed larger and better indexes, AltaVista raised the bar again with a 100 million-page index, unveiled in late 1997.

AltaVista is the place on the Internet in which documents from across the Internet are cataloged—word by word. Want to find all occurrences of "for whom the bell tolls" on the Internet? AltaVista is the place. Want to know which newsgroup postings in the last three days have mentioned your company by name? AltaVista will show you.

With the help of Scooter, a program that roams around the Web to collect Web addresses and Web pages, AltaVista doesn't merely index key terms or ideas, but rather collects complete pages, and the pages they connect to, and the pages those pages connect to, and so on across the Internet. As Scooter collects the pages, they're submitted to the indexing software and made available through AltaVista. Anything that Scooter finds, you can find too, just by typing a word or two into AltaVista. Additionally, you can find new information just about as quickly as it's available—as it's submitted to AltaVista or when Scooter finds it—without having to wait for someone to determine where that new information should fit in a database or catalog. Because Scooter scours the Web constantly, you're more likely to find new or obscure sites with AltaVista than you might with other services.

AltaVista's Speed

Even beyond the technical achievement of indexing the whole Web, AltaVista allows you to search the whole Web remarkably quickly—faster, probably, than you can find something on your desk. (Well, a *lot* quicker than finding something on our desks.)

With the help of sophisticated software, robust Internet connections, and dozens of state-of-the-art Digital Equipment Corporation Alpha servers—called Turbos—you can search through the entire Internet in less time than it usually takes to find a file on a personal computer. If you type in a search term, grab your mouse, and click the Search button, you're likely to have results on your screen by the time your hand returns to the keyboard. Now that's fast!

AltaVista's Ease of Use

All of AltaVista's scope and speed would mean nothing if it weren't easy to use. With AltaVista, you can use a Simple Search, type in words that interest you, and review the results. It's really that easy. And even with that ease, you still take advantage of the full power and scope of AltaVista. (Simple Search is the one the AltaVista developers generally use.) Advanced or power users, or anyone who needs to construct particularly detailed or complex searches, can use AltaVista's Advanced Search mode to develop precise search strategies that isolate carefully chosen pieces of information.

AltaVista's Innovation

Even beyond the other AltaVista advantages, the most impressive aspect of AltaVista is the continuous innovation and search breakthroughs that make it ever easier for you to find the information you need. AltaVista has unveiled advances like Refine, which lets you focus your search on just the areas you're interested in and lets you easily visualize the information resources at your disposal. Additionally, the AltaVista Translation Service, offered in partnership with Systran, makes even resources written in foreign languages available to you to read and use, with only a click of the mouse.

WHAT'S SO REVOLUTIONARY, ANYWAY?

AltaVista solves the wide range of problems associated with finding information in the complexity and broad scope of the Internet. With AltaVista, the difficulties involved with cataloging information, developing databases that selectively include words and phrases, as well as the overriding problems with too much available information, are no more. You no longer need to use multiple search services, indexes, and directories in an effort to find your information. AltaVista allows you to selectively and easily get only the information you need, when you need it.

Pretty cool—but what's so revolutionary about a high-powered index? What does AltaVista do that makes it so special? To give you an idea, step back a few paces and look at how the Internet began, how it grew, and how it created a seemingly unwieldy mass of information (unwieldy, of course, until AltaVista came along).

In the Beginning . . .

The Internet, started in 1969 as a Department of Defense research project, connects millions of computers worldwide in a complex and ever-changing network. Particularly in the past five or six years, with the introduction of PC-based Web browsers like Mosaic and Netscape and the introduction of fast and affordable Internet services, the growth of the Internet has been remarkable. The synergy of these millions of Internet users provided an incredible store of knowledge and a remarkable capacity for collecting and disseminating information.

Unfortunately, the same diversity that gave the Internet its power also made the organization of the Internet remarkably chaotic. The Internet's almost apocalyptic chaos stemmed from unfettered growth combined with little structure or organization. As prices of computer hardware and network infrastructure fell, more people and more companies jumped on the bandwagon, purchased Internet access, and sought out or provided information. The decentralized design of the Internet meant that this process of adding more computers and smaller networks to the Internet was relatively easy. However, the loosely-woven network design also made it almost impossible to track the growth. For example, computers and whole subnetworks could be added or removed with no notice or warning—and often with few people the wiser.

Additionally, information providers flocked to the Internet as it grew. Just as the growth of the Internet was exponential, so too was the rise in these information sources. Many sites, while useful to those who knew about them, often went unnoticed by others on the Internet due to poor communications, perceived low interest, or other human or technical difficulties. Unfortunately, "if you build it, they will come" didn't always apply. The result was a ton of information that was virtually unusable.

Then Came Information Search Problems . . .

As the Internet grew by leaps and bounds, searching for information went from difficult to virtually impossible. For many people, the Internet was an information nightmare because the search tools available were primitive, inadequate for the sheer volume of what the Internet provided. "Information overload" didn't begin to describe the information searching problem that this enormous growth caused on the Web and in Usenet newsgroups.

The World Wide Web

As the Internet—or, more specifically, the World Wide Web—grew, many new pages appeared on servers around the world, but there was no mechanism in place to ensure that someone knew about them. Additionally, when older pages were removed from the network or moved to different places, nobody took care of notifying everyone—or even *someone*—of the changes. Seasoned Internet veterans grew accustomed to information sources disappearing, then reappearing elsewhere, but neither veterans nor novices could consistently seek out vanished information sources or reliable replacements without a growing sense of frustration and helplessness.

Even keeping track of new additions to the Internet overwhelmed most users. To this day, many people dedicate substantial amounts of time keeping track of all the known Internet resources within a particular—increasingly narrowly defined—subject domain. This gloomy picture of the melange of information on the World Wide Web was quite discouraging for information seekers, but remained far better than the searchability of the Usenet discussion groups.

Usenet Newsgroups

Usenet (or network news) discussion groups posed a different set of difficulties for information seekers, including the problems of scope, quantity, and instability. Usenet news grew to its current count of over 20,000 newsgroups, covering literally every topic imaginable. These groups generated millions of messages each day, far beyond the scope of anyone's ability to keep up. Most people couldn't even keep track of the top three or four groups that interested them, let alone the twenty or more that contained information they needed.

As quickly as these Usenet messages appeared, they "expired" and disappeared from the Internet. Usenet messages usually remained accessible for between two to six weeks, at which point they were (and still are) routinely and automatically deleted. Even if someone were to keep up with the new messages as they appeared, the older ones would be disappearing just as quickly.

Finding useful information in Usenet or on the Web required perseverance, patience, quick reactions, and the willingness to plow through a lot of irrelevant or useless data. Early attempts to solve the enormous information problem helped some, but still didn't offer much relief.

Then Came a Few Solutions . . . (Sort Of)

As a result of the growth in Internet users, service providers, available information, and retrieval problems, software engineers and information specialists alike attempted to catalog, index, or otherwise provide improved access to useful sites on the Internet, but with generally limited success. In earlier days, arcane tools—remember Archie, Veronica, and Jughead?—provided information about specific sites on the Internet. Although they were by no means comprehensive, they did offer a starting point to track down data. Skilled Internet users learned to use the existing Internet resources to identify the information they needed.

Out of this mess grew the first attempts to bring order out of chaos. The development initially moved in two different directions—toward *directories* and *search engines*.

Directories

Internet directories provide hierarchically organized lists of links and lend themselves to browsing. Directories, such as Yahoo, present a carefully categorized, apparently logically structured facade to World Wide Web-based information. Laborious manual categorization and indexing provides information seekers easy access to select sites.

Of course, sites that haven't been cataloged can't be found, and with the rate of World Wide Web growth, there isn't a team anywhere that can keep up with cataloging new sites as quickly as they appear. Additionally, manual catalogs are fairly limited in cross-references—it just is not possible to anticipate all of the potentially related topics and to provide pointers to other areas of a catalog to cover all contingencies.

Search Engines

Search engines maintain a database with links to Internet resources and are great for quickly identifying a specific resource. For example, if you are searching for

information about movies, you can swiftly search for any sites that discuss movies without having to wait for an individual to catalog the site.

Search engines collect information about Web sites and allow seekers to search for specific data. The downside is that a searchable database of the whole Internet is enormous; to build an index and present that information to the Internet community requires significant computing power.

As a result, some search engines index only *some* of the words or *some* of the information from *some* of the sites. Indexing more than just the significant words describing a site often poses an overwhelming technological burden for the service providers. Searchers could find anything that's in the database—but, of course, could not find anything that wasn't there.

These early indexing attempts were less than ideal because of the impossibility of anticipating all user needs. While it was possible to index the "significant" words of documents, in addition to titles and descriptions, guessing exactly which words would be useful to future searchers wasn't possible . . . or was it?

Then Came AltaVista . . .

AltaVista applies advanced software technologies, high-bandwidth networking, and the raw computing power of 64-bit Alpha Turbo servers to the immensity of the Internet—it's the classic irresistible-force-meets-immovable-object situation. The Alpha's irresistible force won out, and impressively!

AltaVista became the standard for the Internet—the standard against which all other search engines are measured. The success of AltaVista since its debut in December 1995 was so incredible, even by Internet standards of instant popularity, that Digital has dedicated incredible resources to make the Internet and computing in general easier and more productive for everyone.

In addition to continuing to provide the original comprehensive service at no charge to the Internet community, AltaVista also powers the Web searches of everyone's favorite directory services, including Yahoo, by providing search services and resources when the directories don't contain any matches for an obscure or very specific search. Why? Simply put, because of AltaVista's revolutionary effectiveness.

Over the past couple of years, many more search services have sprung up, offering a variety of tools to help people use the wealth of information on the Internet. AltaVista, meanwhile, has continued to innovate and develop. By introducing new interfaces, new technologies, and other features, like the AltaVista network to allow faster searching from strategically located sites worldwide, AltaVista remains at the head of the pack of Internet search services. Additionally, while many "search"

Getting Started with AltaVista

Don't let the name fool you! Simple Search—the first page you see at AltaVista—is a powerful search tool, enabling you to find any piece of information on the Internet in just a few seconds. "Simple" just means that you can search for information using simple queries—type in a few words or a question, and get the information you need. Plus, you can focus your searches using several different (simple!) techniques to include specific information in the search results or exclude it.

In this chapter, we introduce you to the Simple Search page, show you how to do Simple Searches on the Web, and show you how to hone your searches to get just the information you need. AltaVista always provides comprehensive results, but it takes some practice to focus the results so you get just what you're looking for. This chapter shows you how to find anything on the Internet and how to keep from becoming overwhelmed with information—avoiding the old "drinking from a fire hose" problem.

Tip *Find out how to do Simple Searches of Usenet newsgroups in Chapter 5!*

CONNECTING TO ALTAVISTA

Whether you're planning to do a quick Simple Search or really delve into the nuts and bolts of searching, your first step is to get to the AltaVista home page. Just follow these easy steps:

1. Connect to the Internet. Dial into your Internet service provider (ISP) if you have to—or just get a cup of coffee and rejoice in being able to use your company's Internet connection so freely!
2. Open your browser.
3. Connect to AltaVista by typing **http://altavista.digital.com** in the Location field of your browser.

Note *Be sure to use altavista.**digital**.com in the URL. Many people assume that altavista.com is the AltaVista Search Service, but it's actually a different company with no connection to Digital Equipment Corporation or the AltaVista Search Service. This company does provide the search form as a convenience to their visitors (you'll find information about adding the form to a Web page in Chapter 6), but the best results and complete AltaVista features are only available through altavista.digital.com.*

4. Press ENTER.

You're here! Welcome to AltaVista! The AltaVista home page provides you with all the tools you'll need to conduct Simple Searches and to navigate to other parts of the AltaVista service. Figure 2-1 shows you the AltaVista home page features, and the table following describes what each piece of the page does.

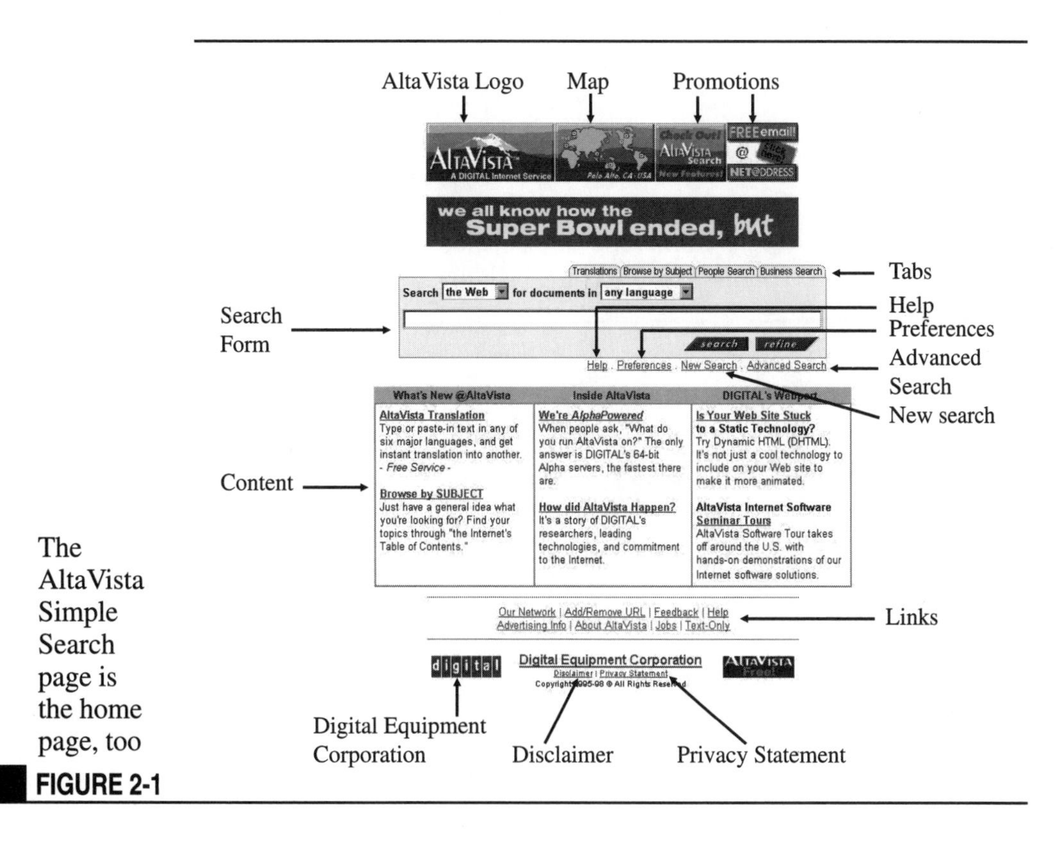

The AltaVista Simple Search page is the home page, too

FIGURE 2-1

Note *If you choose text-only mode, you will not see the graphics. You will see links to each of the different parts of AltaVista, though.*

AltaVista Logo (at the left)	Takes you back to the AltaVista home page.
Map	Links to information about the AltaVista Search Network.
Promotions	Fills out the rest of the navigation bar. These ads, usually for special or new parts of the AltaVista service, change all the time. You can click on the ads to find out more.
Tabs	Lets you access other parts of the AltaVista Search Service. We cover Translations later in this chapter, and Browse by Subject, People Search, and Business Search in Chapter 3.
Search Form	Lets you choose what to search and how to present the results. It also gives you a place to input your search criteria. We'll tell you more about this form later in this chapter.
Help	Whisks you off to—you guessed it—help. There's good information here for those willing to read the instructions. Of course, if you're reading this book, you're getting help in spades already.
Preferences	Lets you customize AltaVista to work the way you do. More about this in Chapter 7.
New Search	Takes you back to the home page or, for practical purposes, to the Simple Search location.
Advanced Search	Takes you to the Advanced Search page. We cover this in Chapter 5.
Content	Provides brief information about and links to AltaVista news, announcements, features, or special offerings.
Our Network	Links to information about the AltaVista search network. You'll find more on this in Chapter 10.
Add/Remove URL	Lets you submit a specific URL to AltaVista. Find out more on this in Chapter 8.

Feedback	Lets you contact the AltaVista team to offer comments or other feedback.
Help	Takes you to the online help pages (same as the Help link on previous page), but if you're reading this book, you're not likely to need help for anything other than a memory jog.
Advertising Info	Provides everything you need to put your ads on AltaVista.
About AltaVista	Provides the latest and greatest news about changes and developments on AltaVista, including links to press releases and other announcements.
Jobs	Lists information about employment opportunities at AltaVista.
Text-Only	Provides an interface optimized for older or non-graphical browsers. If you're using one of these browsers, AltaVista will likely automatically present you with the Text-Only page. At any rate, if you're using a text-only browser, like Lynx, the Text-Only page is a better starting point than the regular Simple Search page.
Digital Equipment Corporation	Takes you to the company that makes the computers that make AltaVista run.
Disclaimer	Provides some information about use of the AltaVista logo and the rights that you do—and don't—have to use AltaVista and the AltaVista logo. Nothing nasty lurks here, unless you're reselling AltaVista services as if they were your own.
Privacy Statement	Gives you the information you need about the steps AltaVista takes to safeguard your privacy when searching the Net.

DOING A SIMPLE SEARCH

Ready to get started? Okay, let's go! Take a look at AltaVista's Simple Search form, as shown in Figure 2-2.

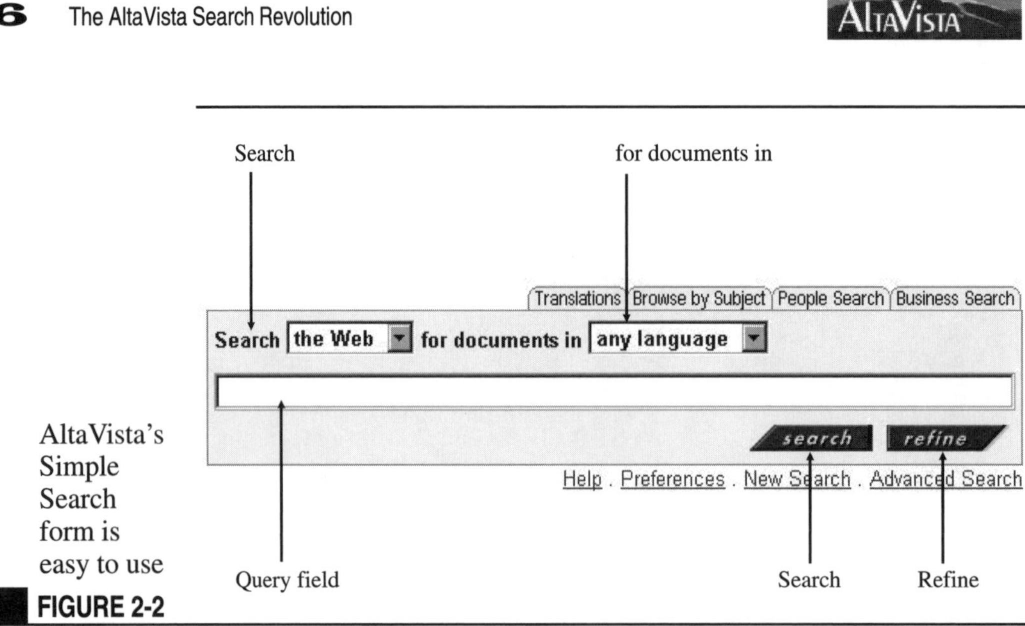

AltaVista's Simple Search form is easy to use

FIGURE 2-2

Each part of the form has a specific function, as described here:

Search	The drop-down Search menu allows you to choose whether you want to search the Web or Usenet newsgroups. This chapter focuses on the more common Web searches, while Chapter 6 contains the full scoop on Usenet searches.
for documents in	With this drop-down menu, you can specify which language you want to find results in.
Query field	Provides space for you to type in the search criteria, including significant words and any other pertinent information. Your search criteria will be in this field even after you submit your query, so you can easily edit and revise the search criteria.
Search	Click this button to submit your query. (You may also be able to submit the query simply by pressing ENTER, depending on what type of browser you're using.)
Refine	Whisks you to AltaVista's Refine page (assuming you've already entered search criteria). On the Refine page you can easily narrow and precisely target your search results. You'll find more about Refine in Chapter 4.

So, armed with this information, let's take a look at doing Simple Searches on the Web. Basically, all you have to do is fill in the Query field with the word or words you're searching for. In fact, because AltaVista's Simple Search is so easy, you can simply enter the words phrased as a question, or however they happen to pop into your mind. For this example, let's suppose we're looking for information about tornadoes.

1. Leave the default menu selections for *Search* and *for documents in* as they stand. This default method of searching looks through the Web and selects documents in any language, rather than restricting the language to just English, for example.
2. Click in the Query field (that blank area just begging to be filled with your search terms).
3. Type the word or words you want to search for. For very basic Simple Searches, you can type in:
   - a single word (such as **tornado**)
   - a group of related terms (such as **tornado wind damage Oklahoma**)
   - a natural-language query (such as **how much damage did tornadoes do in 1997?** or **tell me about 1998 tornado damage**)
4. Click Search (or press ENTER, depending on your browser).

That's it! Just by entering a word and clicking Search, you get loads of useful matches (also called hits, results, or listings, depending on who you're talking to).

UNDERSTANDING SEARCH RESULTS

Anytime you do a general search, such as a search for single words or phrases, you're likely to get a lot of hits—thousands, sometimes millions. Don't be alarmed or discouraged about a high number of results, though. No matter what your search query, AltaVista presents the results with what it thinks are the most likely matches at the top of the list. Even if the results pages seem like they go on and on, you'll likely find the best matches within the first 10 or 20 hits. If you don't find what you need within the first 30 or 40 matches, you should probably work on improving your query. You certainly don't have to wade through 150 matches to find what you need—AltaVista does the hard work for you. Whew!

As a rule, if you don't think that the results AltaVista provides are right, you probably need to provide more information for the service to work with. For

example, if you search for *elk*, you will get any documents in which the word "elk" figures prominently. If you're looking specifically for information about *feeding* elk, you need to provide AltaVista with that information—as in *elk feeding*.

Simple Search results come in two varieties: Standard and Compact. Each format has its own advantages, as discussed in the next two sections.

Understanding Search Form Results

By default, AltaVista Web searches give a detailed view of the results list. This default selection is a great choice when you need details about the search results to help you determine whether they're relevant to your needs. Let's take a look at the results of the **tornado damage** search example, shown in Figure 2-3.

Not only does the Detailed (default) results format provide links to sites, but it also provides much more information, as summarized in the following table.

Results Page Element	Description
Approximate number of matches	Indicates approximately how many documents AltaVista found that match your criteria. This number will vary somewhat, depending on how busy the server is; AltaVista won't count all the matches if it would slow down other searches, so you might get just an estimate.
The title of each item	Links directly to the document. This "title" might not appear in the body of the document, but it is the title that the author has given to the document. If the author did not provide a title, in place of the title will be "No Title."
Text from the document	Gives you a brief idea of what to expect. This text is either a specific description provided by the author, or just the first text in the document. See Chapter 8 for more information about specifying text for AltaVista.
The Web address or URL	Takes you to the document. It can also provide useful information about what the document contains (for example, instructions.html or copyright.htm) and who published the document (for example, digital.com, harvard.edu, navy.mil).

Results Page Element	Description
The size of the text within the file	Shows how much space the actual text of the document requires. If you're looking for a thorough article on a specific topic, you'll probably want to look for pages with 10K or more—smaller pages, of 1 or 2K, may not provide all the information you need. Some pages with little text might have plentiful graphics, so this size offers only a rough guideline.
Date the file was last modified	Helps you determine how relevant the information is to your needs, based on how current it is.
Language	Tells you what language the body of the document is in. Occasionally the title is in English, but the body of the document is in another language, so AltaVista goes with the most prevalent language in the document.
Translate	Links to the AltaVista Translation Service, so you can read documents even if you don't speak the language in which they are written.
Page numbers of the results	Lets you move forward or backward through the results pages. The yellow dot shows you which page of the results you're on. You can select a page by clicking on the number button or clicking Next (or Previous) to move back and forth through the pages. The number under Next (or Previous) tells you what the next page will be.
Word count	Tells how many times each of your search terms was found. Remember, don't be overwhelmed with the word count; AltaVista places the best matches at the top of the results list, so you don't have to wade through everything.
Ignored	Lists your search terms that were too common to be useful (if any) and that are therefore not represented in the results.
Tip	Provides hints and tricks about the most effective ways to use AltaVista.

Translations | Browse by Subject | People Search | Business Search

Search the Web for documents in any language

tornado damage

search refine

Help . Preferences . New Search . Advanced Search

Approximate number of matches → About 30126 documents match your query.

Click to find related books at amazon.com

1. Tornado Damage Assessment in VORTEX-95
Tornado Damage Assessment in VORTEX-95. Erik N. Rasmussen. Casey Crosbie. Cooperative Institute for Mesoscale Meteorological Studies and National Severe...
http://doplight.nssl.noaa.gov/~erik/ConfPapers/VORTEX/damage/tornadoTracks.html - size 21K - 1-Feb-96 - English - Translate

Title → 2. Tornado Damage Galleries
Tornado Damage Galleries. These are photos of tornado damage I have taken over the course of my research study. ← Text
You may use the photos within these pages..
Web address or URL → http://www.geocities.com/CapeCanaveral/Lab/2430/torgal.html - size 1K - 1-Oct-97 - English - Translate

3. Oakfield Tornado Damage Photo 6
type description here
http://www.theforge.com/tornado/past/oakdam6.htm - size 2K - 9-Apr-97 - English - Translate

Size of file

4. Oakfield Tornado Damage Photo 7
type description here
http://www.theforge.com/tornado/past/oakdam7.htm - size 2K - 9-Apr-97 - English - Translate

Date of file

5. Hope, Arkansas: Tornado Damage
Hope and Hempstead County Tornado Damage. The tornado that claimed the lives of twenty-four Arkansans and injured hundreds more on Saturday, March first,..
http://www.aiea.ualr.edu/dina/cities/hope/tornado.html - size 3K - 20-Aug-97 - English - Translate

Language

6. Arkadelphia, Arkansas: Arkadelphia and Clark County Tornado Damage
Arkadelphia and Clark County Tornado Damage. Photograph courtesy of the Arkansas Democrat-Gazette. On Saturday, March first, the first severe weather of...
http://www.aiea.ualr.edu/dina/cities/arkadelphia/storm.html - size 4K - 28-Aug-97 - English - Translate

Translate

7. Jarrell, Texas Tornado Damage
Last revised: July 27, 1997. Damage from the May 27, 1997 Jarrell, Texas Tornado. The extreme violence unleashed on the small town of Jarrell, Texas by...
http://www.pulse.net/storm/jarrell.htm - size 16K - 9-Sep-97 - English - Translate

8. Berea Tornado Damage April 20, 1996
Berea Tornado Damage April 20, 1996. The storm which ripped through Berea on April 20 had winds gusting at 120mph. The following photos show some of the...
http://www.4berea.com/storm/ - size 2K - 5-Sep-97 - English - Translate

9. Oakfield Tornado Damage Photo 2
type description here
http://www.theforge.com/tornado/past/oakdam2.htm - size 2K - 9-Apr-97 - English - Translate

10. Despite Tornado Damage, Focus: HOPE is Up and Running!
July 6, 1997 - Despite Tornado Damage, Focus: HOPE is Up and Running! Eleanor Josaitis, Focus: HOPE executive director, has had the full cooperation of...
http://www.focushope.edu/legend/news.events/1997/tornado/tornado.html - size 4K - 11-Jul-97 - English - Translate

Tip: If the first few results don't satisfy you and there are many results, try the **refine** button. ← Tip

Word count: tornado: 245687; damage: 1949275

↑ Word count

1 2 3 4 5 6 7 8 9 10 11 12 13 14 15 16 17 18 19 20 2 ← Page numbers

AltaVista's default results provide lots of details

FIGURE 2-3

Understanding Compact Form Results

If you prefer shorter and more concise search results, you can try the *Compact Form,* which provides only a fragment of the information that the Detailed Form provides. The information Compact Form provides is very similar to that of Detailed—some of the data is truncated or omitted, but you get all the basics, as shown in Figure 2-4. Compact Form is a good choice if you're sure you can identify the document you

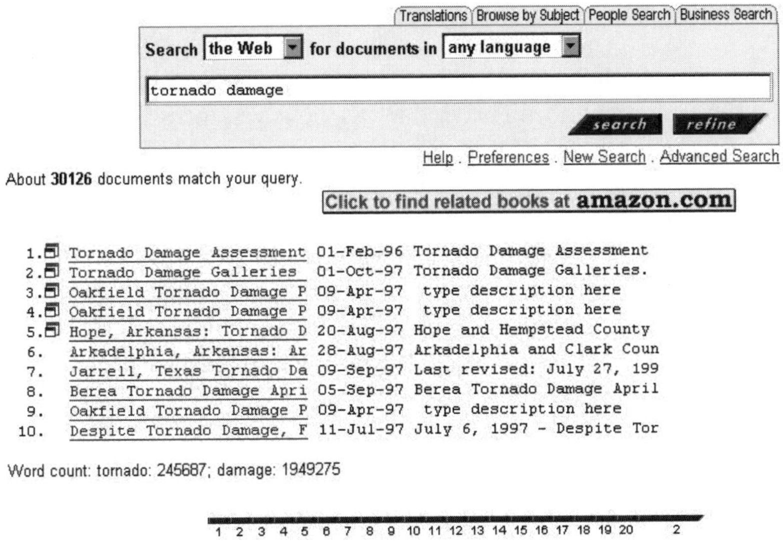

Translations Browse by Subject People Search Business Search

Search the Web for documents in any language

tornado damage

search refine

Help . Preferences . New Search . Advanced Search

About **30126** documents match your query.

Click to find related books at **amazon.com**

```
1.  Tornado Damage Assessment 01-Feb-96 Tornado Damage Assessment
2.  Tornado Damage Galleries  01-Oct-97 Tornado Damage Galleries.
3.  Oakfield Tornado Damage P 09-Apr-97  type description here
4.  Oakfield Tornado Damage P 09-Apr-97  type description here
5.  Hope, Arkansas: Tornado D 20-Aug-97 Hope and Hempstead County
6.  Arkadelphia, Arkansas: Ar 28-Aug-97 Arkadelphia and Clark Coun
7.  Jarrell, Texas Tornado Da 09-Sep-97 Last revised: July 27, 199
8.  Berea Tornado Damage Apri 05-Sep-97 Berea Tornado Damage April
9.  Oakfield Tornado Damage P 09-Apr-97  type description here
10. Despite Tornado Damage, F 11-Jul-97 July 6, 1997 - Despite Tor
```

Word count: tornado: 245687; damage: 1949275

1 2 3 4 5 6 7 8 9 10 11 12 13 14 15 16 17 18 19 20 2

Compact Form results give you the basics

FIGURE 2-4

are looking for with very little information. (Heavy users of Compact Form tend to be pretty good at "Name That Tune" as well.)

As you can see, Compact Form is really just a truncated version of Detailed Form. You get the same categories of information, but you get fewer details within the categories. To switch to Compact Form, follow these quick steps:

1. Click Preferences (below the Search button).
2. Check Compact Form for Web Results.
3. Click the Set Preferences button at the bottom of the Preferences page. AltaVista takes you back to the Simple Search page.

That's it! With this quick change, you'll get your results in Compact Form. To switch back to Standard Form, go back to Preferences and Set Preferences without selecting the Compact Form option. You'll return to the main AltaVista page with the Standard Form results.

Tip *See Chapter 7 for more information about specifying Preferences.*

HONING YOUR SIMPLE SEARCH SKILLS

As we've shown you throughout this chapter, doing Simple Searches is truly—well—simple. Just by typing in a word, a group of words, or a question, you can find the information you need. You can, however, focus your searches precisely and use—and reuse—the results in creative and valuable ways. Doing so allows you to take advantage of the full resources of the Internet to meet your business and personal needs, and cuts way down on the time you spend looking at irrelevant results.

While there are plenty of additional techniques and tricks that make searches with AltaVista even easier (we'll be giving you specifics throughout the book), the techniques discussed in the following sections give you a good start. We recommend that you try out these tips as you're working through this chapter so that you can see how each improves search results:

- Using rare words
- Using all words that might matter
- Using phrases when you know the exact word order
- Using punctuation and spaces effectively
- Using capitalization effectively
- Searching according to fields
- Requiring specific words or phrases
- Excluding specific words or phrases
- Combining elements in your searches
- Using wildcards when you aren't sure of the exact word

Using Rare Words

Using rare words—that is, unusual names or terms—is a great way to improve your search results. The more precise you can be about what you are looking for by using rare terms, the better your results will be. For example, use terms like *collie* instead of *dog* or *pet*; *Dilbert* instead of *comic strip*; or *hyperactivity* rather than *behavior*. In doing so, you're more likely to find the exact piece of information you need.

Remember *How common or rare a word is does not depend on general usage but rather on what actually appears on the World Wide Web.*

Using rare words in your search is important because of the way AltaVista works. When you search for a specific word, AltaVista checks through an index that contains every word found on over 100 million Web pages that reside on over 1 million servers around the world. If the word you asked for is common, it has probably appeared many, many times (perhaps thousands or millions of times!). For instance, if you search for the word *music,* you'll find over a million documents. You've now gained a piece of trivia, but you aren't likely to be able to scan through millions of documents to isolate the ones with the information you want. Entering a more specific word, such as *musicals,* gives you about 10,000 documents—a somewhat more manageable number, but still pretty hard to use. Getting as specific as possible and entering the rare word *Evita* gives you more precisely focused selections—focused enough that you can easily scan through the first few pages of matches and likely find the information you need.

Note *Don't be so specific that you exclude potentially valid pages. If you want information about garden shade trees but search for* magnolia *because you like magnolia trees and magnolia gives you a manageable number of matches, you're not likely to find information about similar and appropriate trees.*

Using All Words That Might Matter

As mentioned earlier in this chapter, you can do a Simple Search just by entering a series of words or by entering a question. The point here, though, is to be sure you enter all the words that might affect the results. Adding a pertinent word (or leaving one out) can significantly change the order in which the results are listed, which affects how useful the results are.

For example, suppose you want to know the population of Alaska's capitol. While you could enter the search query **capitol Alaska** or **Alaska capitol**, a better bet would be to search for **capitol Alaska population**. If you do so, documents with information about Juneau's population would appear higher in the results list than if you don't include the word population in the search query. (Of course, if you happen to know that Juneau is the capitol, that'd be a good term to add as well.)

Or, you could enter this same basic query as a question: **What is the population of the capitol of Alaska?** AltaVista searches for information using the rare words

in the question—in this case, *capitol*, *Alaska*, and *population*—and ignores common words such as *what*, *is*, *the*, and *of*. In fact, AltaVista doesn't really know it's a question at all; your query is just processed as a bunch of words. It'll search for the rare or uncommon words and ignore the rest.

Using Phrases When You Know the Exact Word Order

Another way to improve your search efforts is to create exact search phrases by enclosing words inside quotation marks (""). A *phrase* is any set of words that appears in a specific order. Using a Simple Search, which looks for each word individually, is not the best way to find an exact phrase.

For example, if you entered the words *to be or not to be,* you would get no results at all because the words are all so common. Each of those words appears tens of millions of times—too frequently to be of any use. If you entered those same words using quotation marks, however, you would get over 8,000 matches—all of them Web pages that include that complete phrase. You could narrow the search even further by including more words within the quotation marks, such as *"to be or not to be, that is the question,"* which will lead you straight to 1,000 or so Web pages, many of which are about Shakespeare's works (and many others illustrating that reusing quotations from Shakespeare isn't as unique or clever as some people think).

While some other search engines discard common English words like *a* and *the* and don't keep track of complete phrases like *to be or not to be* to conserve space, AltaVista saves every one of them. That's what makes it possible to do exact searches for complete phrases and sentences. Used in combination with other words and commands, searching for exact phrases is a valuable tool for narrowing searches, as well as for looking up quotations.

Using Punctuation and Spaces Effectively

Another way to improve your search is to use punctuation and spaces carefully. AltaVista indexes only words—not spaces or punctuation marks. AltaVista considers a *word* to be any group of letters and digits, while *spaces* indicate the end of one word and the beginning of the next.

When you type in a query using just words separated by spaces, AltaVista treats those words as separate entities, and their order makes no difference. When you include punctuation in your query, you also link those words together in a specific order to form phrases. That means that words connected with any punctuation (such as a comma, period, colon, or semicolon) are treated as phrases—just as they would be if the same string of words were included inside quotation marks, as discussed in the previous section.

AltaVista's approach to punctuation has several benefits. For instance, many people are inconsistent in their use of hyphens, slashes, and other characters in product names and model numbers. When you use AltaVista to find such terms, you don't have to try to imagine all the possible variants of punctuation and search for all of them. You can enter one of them, and the results will include all of them, giving them all equal weight. For instance, AltaVista would treat *Hale-Bopp,* and *"Hale Bopp"* (as in the comet) identically. *HaleBopp*, however, is different—it's one word, not two. To find all possible variants, you might search for *Hale-Bopp HaleBopp.*

Using Capitalization Effectively

In addition to using punctuation and spaces carefully, you should think about how you can use capitalization. AltaVista is case-sensitive, but only when you want it to be. When it indexes text, it preserves the fact that certain letters are capitalized and others are lowercase. The word *RED* is indexed differently and, therefore, shows up differently from the words *Red* or *red.*

If you enter your query with all lowercase letters, you will get matches for all instances of those words, regardless of capitalization. For example, if you enter *red*, AltaVista will return hits for pages with *RED*, *Red*, *rEd*, *reD*, and so on. On the other hand, if you include capitalization in your query, you will only get hits that exactly match that pattern of capitalization. *RED* will only give you matches for *RED*—not for *red* or any other variant.

This approach has two important benefits. First, you can easily search for trade names where the word may be common but the capitalization is unique, such as *AltaVista* or *WiReD.*

You can also easily distinguish between proper nouns and ordinary nouns, such as *Turkey* the country and *turkey* the bird, *Frank* the name and *frank* the adjective, and *Who* the rock group and *who* the pronoun (although the pronoun *who,* if capitalized at the beginning of a sentence, will show up on a *Who* search as well). Second, you can easily search for all variations of capitalization of a particular word by entering a single query in lowercase, rather than having to enter all possible permutations.

Tip *If you are not sure about the capitalization of a word, use all lowercase letters. And if you* are *sure, include the capital letters to help narrow your search.*

Searching According to Fields

You can help improve your search results by searching according to *fields*, which are just parts of a Web page, such as a title, a link, or an image. (The term *fields* comes from database terminology—when searching the Web by these specific parts, it's much like searching a database for a specific field.)

Searching by field can lead to particularly focused results. For example, suppose you're searching for information about Elvis. You could do a Simple Search and find results that include the word *Elvis* in any field. Searching by field, however, you can search for documents that include the word *Elvis* in one particular part of a Web page. In doing so, you are more likely to get a list of documents with the specific content you need, rather than a list that includes all documents with any mention of him.

Tip *You can also search newsgroup postings by field. Check out Chapter 6 for more information.*

AltaVista lets you search for ten kinds of Web page fields: text, title, link, anchor, url, host, domain, image, applet, and object. Constructing a field query is a cinch. All you have to do is enter the field name, follow it with a colon, and then include the word or phrase, like this: *title:"Elvis sightings."* This query would find documents with the words *Elvis sightings* in the document title.

Tip *Other search fields might be added in the future. Check the online Help file for updates.*

The following sections discuss each of the fields in detail. Before you visit those sections, breeze through the following table, which summarizes Web fields and gives you examples of queries.

Field	Description	Sample Query
text:	Restricts the search to the body of the document.	text:"Elvis sightings"
title:	Searches only for the part of the document that the creator explicitly labeled as a title.	title:"Blue Suede Shoes"
link:	Searches only for a hypertext link (the URL) embedded in the document.	link:elvis.com

Field	Description	Sample Query
anchor:	Searches only for the visible part of a hypertext link (the words you click on).	anchor:Elvis
url:	Searches for pages that have these words as part of their address (URL).	url:elvis.fan.com
host:	Searches only on the computer name of the system where the pages reside.	host:elvis.com
domain:	Searches for the domain in the URL.	domain:com or domain:us
image:	Searches for the name of an image or picture.	image:elvis.gif
applet:	Searches only for the names of applets.	applet:king
object:	Searches only for the names of objects.	object:king

text:

Searching by *text:* finds information located in the document text. Not links, not Web addresses—just plain text, headings, and other content. With this type of field query you could, for example, search for pages that discuss *altavista.digital.com*, rather than all of the pages that link to it.

title:

Searching by *title:* provides information located in the document title—the title you would see in your browser's title bar. These searches are handy when you are trying to find specific words or phrases in the title or when you know a standard naming convention. For instance, *title:"Fab Four"* would bring up any documents with the phrase *Fab Four* in the title. Keep in mind, though, that this example would not find any documents with *Fab Four* in the body of the document but not in the title, so a Web page entitled "Beatles" would not come up.

link: and anchor:

AltaVista also lets you search for *link:* and *anchor*: fields. These hyperlink fields are what let you leap from page to page on the Web. Specifically, the anchor itself is the text you click on; the link is the Web address that you actually jump to.

Suppose you're looking for information about the White House. You want to see pages that link to or refer to the White House. You could do a field search for **link://www.whitehouse.gov**. This search would yield documents that include a link to this White House URL. Or, you could do the field search **anchor:"White House"** to find documents that use the words *White House* as the link text or the ALT= text of a linked image.

Finding pages by link allows you to identify approximately how many documents link to a certain page, which can be very useful to Webmasters. Additionally, the *link:* field lets you find all links regardless of how the author identified the anchor in the text. Remember that a lot of links in Web pages do not display the URL; usually you'll just see descriptive text instead of the actual URL of the page you're linking to.

Remember *If you're looking for a URL, use* link:*; if you're looking for the actual text that people would click, use* anchor:.

url:, host:, and domain:

The fields *host:*, *url:*, and *domain:* each look at different parts of a Web address:

- *url:* looks at the complete address—the domain name as well as any directory or filenames that follow it. A search for **url:modems.html** would look for pages that have that filename, perhaps after some directory names and the host name.
- *host*: looks at the name of the computer where the pages reside. For example, a search for **host:yoursite.com** would show all pages from yoursite.com that AltaVista has indexed, including pages from mail.yoursite.com or hobbes.yoursite.com.
- *domain:* looks at the domain from which the page comes, like *.de* for Germany or *.uk* for the United Kingdom, or *.edu* for educational institutions in the United States.

Domains

Every country on the Internet has its own domain, and several other domains exist for use primarily in the United States. Some of the most common domains you might encounter are listed in the tables here. The first set of domains is primarily found in the United States, although certainly not exclusively so.

.com	For commercial enterprises.
.edu	For educational institutions.
.gov	For governmental agencies.
.mil	For defense and military computers in the U.S.
.net	For networks and the computers that help to run the Internet.
.org	For nonprofit organizations.

Some international domains (indicating specific countries):

.au	Australia
.ca	Canada
.ch	Switzerland
.de	Germany
.dk	Denmark
.fi	Finland
.fr	France
.it	Italy
.jp	Japan
.uk	United Kingdom
.us	United States
.za	South Africa

image:

The field *image:* looks for the address or name of an image, and searches for an *image:* field name can be a valuable tool for finding documents (or images) with the information you need.

For example, a search for **image:pluto** would return the list of Web pages that have pictures named for the planet, the Roman god, or the Disney cartoon character. Similarly, a search for **image:comet** would match everything from a file named comet.gif to pictures of a famous comet called hale-bopp-comet.gif or comethalley.jpg.

Keep in mind that many images have cryptic names and are unlikely to show up in *image:* searches. For example, a picture of the Hale-Bopp comet called HB1025.GIF would not show up in these searches. You can, though, use the *image:* field to find examples of image files of a particular type. For instance, **image:gif** matches millions of Web pages that include GIF images.

applet:

The field *applet:* returns matches that have Java *applets*—small programs embedded in an HTML document. If you know part or all of the applet's name, you can search using *applet:*—for example, *applet:game.* If you find a really great applet and are considering using it on your home page, you could search using the format *applet:thewholefilename* to see what sites are already using it.

Note *If you are interested in finding samples of other kinds of multimedia files, your best bet would be to do a structured search for files that include the usual file extension name in the address. In other words, search for* url:qt *or* url:avi *for QuickTime and AVI movie files, respectively.*

object:

The field *object:* returns matches that have embedded *objects*—small programs or multimedia files or other non-HTML content embedded in an HTML document. If you know part or all of the applet's name, you can search using *object:*—for example, *object:game.* If you find a really great applet and are considering using it on your home page, you could search using the format *applet:thewholefilename* to see what sites are already using it.

Requiring Specific Words or Phrases

Another way to improve your search results is to require certain words or phrases. Remember that queries with two or more words produce results that contain any of the words requested. For example, the query *left-handed politician* would give a count of all documents that have either word in them, which is a much higher number than the number of documents that contain both those words.

You can tell AltaVista that a particular word must be present by putting a plus sign (+) before the search word, such as the query *+left-handed politician.* This query would show all pages that have the word *left-handed,* and pages with the word *politician* as well would appear at the top of the list. The phrase *+left-handed +politician* would match only pages that included both words (but not necessarily in that order, or even near one another).

You might also want to use the plus sign to affect the ranking of documents. For instance, if you search for *albatross boat fishing,* you'll find that documents that have several instances of *albatross* (a relatively rare word) near the beginning could come out higher on the list than documents that had all three search words in them. By entering instead *+albatross +boat +fishing,* you can ensure that all three words are present in each of the matches.

Note *When requiring or excluding terms, be sure that you have no spaces between the + or - and the term following.*

Excluding Specific Words or Phrases

Just as you can require particular words you want to search for, you can also exclude words that you don't want to see on the pages listed as matches to your search. Excluding words is handy if you know there is more than one way to understand the term you are searching for and you don't want to be inundated with matches that have the right term but the wrong meaning.

For example, if you wanted to find documents about a person named Mona Lisa rather than about the famous painting by that name, you would enter **"Mona Lisa" -"Louvre"** or **"Mona Lisa"-"da Vinci."** AltaVista would return all documents that had the two words *Mona* and *Lisa* together (capitalized as you entered them), but exclude any document that also had the word *Louvre* or the name *da Vinci*. Similarly, *+digital audio recording -"Digital Equipment"* would look for documents that must have the word *digital*, would prioritize those that also had both *audio* and *recording*, and would exclude those that talked about *Digital Equipment* Corporation.

Tip *See Chapter 4 for much more information about excluding topics from your results list.*

Combining Elements in Your Searches

Another way to focus your searches is to combine different terms, phrases, fields, and include (+) and exclude (-) flags. For instance, if you want to know how many Web pages outside of your own site have hypertext links to your site, then type **+link:yourdomainname.com -host:yourdomainname.com.** To find documents about microprocessors, but only at Intel's Web site, enter **+microprocessor +host:intel.com.** AltaVista will return all documents that include the word *microprocessor* and that are located at a server owned by Intel.

Tip *You can use the minus sign to exclude any unwanted results from your search. For example, if you're searching for parenting tips, but you're not interested in issues about stepchildren, just search for* ***"parenting tips" -step*** *and you're there.*

Or, you can combine phrases and fields by enclosing them in quotes (""). For instance, **text:"Hale-Bopp comet" image:.jpg nasa** would look for Web pages with the phrase *Hale Bopp comet* in the text, images in JPEG format, and the word *nasa* anywhere in the document.

Using Wildcards When You Aren't Sure of the Exact Word

So far in this section, we've shown you how to narrow your searches to help you pinpoint the information you're looking for. You may, however, find occasion to broaden your search—for instance, if you're not getting enough hits. You can broaden your search by using *wildcards,* which are simply placeholders for missing letters. Basically, you can replace missing characters or missing words with an asterisk (*) to get hits on any terms that include variants on the base word. For example, if you search for **colo*r** you'll get matches for both the British spelling *colour* and the American *color.*

Because wildcards can broaden your search considerably (and return an unwieldy number of matches), keep in mind these guidelines:

- Use wildcards only after three or more characters.
- Use wildcards as a placeholder for up to five unknown characters.

If you keep within these guidelines, you'll find wildcards wildly useful for broadening your searches. For example, you can use wildcards to catch the plural as well as the singular form of a noun, or different tenses of the same verb, or other grammatical variants and compounds.

You can also use wildcards within a phrase, which is particularly useful when searching for a quotation when there's a word you are unsure of. For instance, in response to the query **"one if * land,"** AltaVista will check for phrases in which the first two words are *one if* and the last word is *land* and only one word separates the beginning from the end. Hence it would match *one if by land*, *one if over land*, *one if on land,* and so forth. This approach could help you determine the correct quotation, if that's your objective, or it could also help you find instances where the author, intentionally or not, reworded the quotation.

DEVELOPING MULTINATIONAL SEARCHES

One of the driving strategies behind AltaVista is to fully support the multinational character of the Web, including different languages and different typographic needs. You can tell AltaVista to retrieve documents published in a language you specify, search for certain characters (like accented or umlauted characters) from multinational documents, and even let AltaVista translate Web pages into a language of your choice. Piqued your curiosity? Let's take a look…

Specifying Languages

As you can see in the following graphic, there's one final component of the Simple Search form—the language selection drop-down menu. By default, AltaVista retrieves documents that match your search queries, regardless of which language they are written in. You can, however, specify that it return only documents published in the language you specify. AltaVista actually looks at the content to determine the language a document is in, so this selector differs significantly from a simple country selector. This capability can be useful both for regular searches—to restrict the results to only the language you speak—and as a language learning aid. If you're studying German, you can require only matches in German to get some extra practice.

Translations Browse by Subject People Search Business Search

Search the Web for documents in any language

any language
Chinese
Czech
Danish
Dutch
English
Estonian
Finnish
French
German
Greek
Hebrew
Hungarian
Icelandic
Italian
Japanese
Korean
Latvian
Lithuanian
Norwegian

search refine

He... ...w Search . Advanced Search

Click the down arrow to display the language drop-down menu, and select the language of your choice. Then, simply enter the search query, and AltaVista will find only matching documents published in the language you specify.

Tip *You can also choose to view documents published in multiple languages. We'll show you how to set these preferences in Chapter 7.*

What Languages Does AltaVista Support?

At press time, AltaVista supported the following languages. Look for new languages to be added in the near future!

Chinese	German	Latvian
Czech	Greek	Lithuanian
Danish	Hebrew	Norwegian
Dutch	Hungarian	Polish
English	Icelandic	Portuguese
Estonian	Italian	Romanian
Finnish	Japanese	Russian
French	Korean	Spanish
		Swedish

Finding Multinational Documents

You can also find documents containing multinational terms. In particular, you can use English letters to stand in for many non-English characters. For example, suppose you're searching for a word that, in its native language, would have an accent, umlaut, slash through it, or other diacritical mark. All you have to do is type in the word without the diacritical marks. So, rather than typing **Tübingen** (the proper spelling of the small university town in Germany), you could just type **Tubingen**. These character substitutions apply to *a, c, e, i, n, o, u,* and *y*, in upper- and lowercase, as summarized in the following table:

Original Characters	Substitution
Æ	AE
Á Â À Å Ã Ä	A
Ç	C
Ð	D
É Ê È Ë	E
Í Î Ì Ï	I
Ñ	N
Ó Ô Ò Ø Õ Ö	O
Þ	TH
Ú Û Ù Ü	U
Ý	Y
Æ	ae
á â à å ã ä	a
ç	c
é ê è ë	e
ð	d
í î ì ï	I
ñ	n
ó ô ò ø õ ö	o
ß	ss
þ	th
ú û ù ü	u
ý	y

You can, however, include accent marks (´) in searches to help narrow your search. Using an accented word in a query forces an exact match on the entire word, while an unaccented word matches with all accented and unaccented variants. Use of accents is similar to use of capitalization—unaccented or lowercase terms match everything, while accented or capitalized words only match like words.

For example, if you use *Eléphant* in a query, you will match only the French spelling for the pachyderm. If you do not enter accents in the search window (and it could be difficult for you to enter them, depending on your browser, keyboard, and computer system), you can always leave off the accents, thereby matching both the French and English spellings.

Note *Many documents on the Web do not include the correct accents—perhaps the documents were created by people who did not have access to a keyboard with all of the needed characters—so a search for accented words might not always find the words you need.*

Translating Documents

One of the newest features AltaVista offers is on-the-fly translation. Really! Say you want to find out exactly what German Web pages have to say about your new product. Just search for the product name and restrict the search to find only documents in German. When you get the results back, you'll see a Translate link at the end of every match.

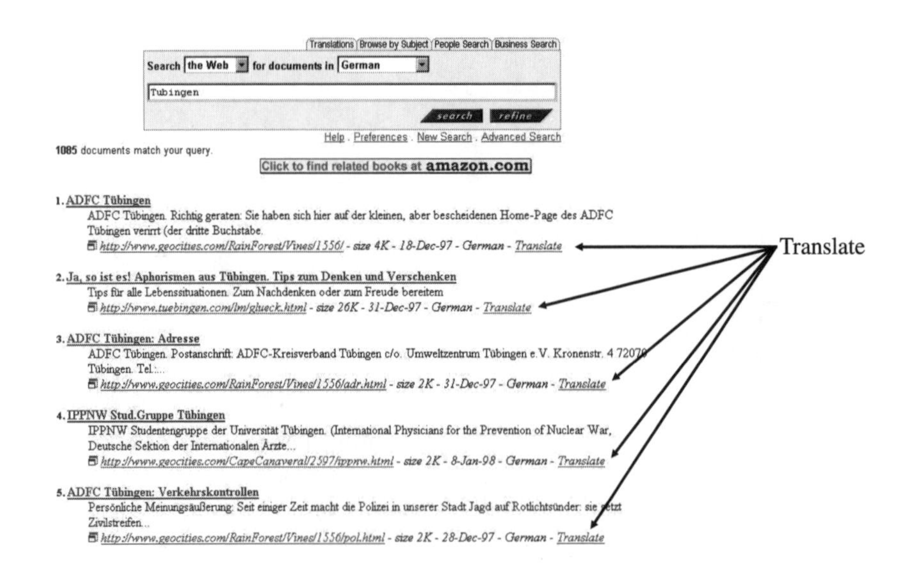

Click any Translate link to bring up the AltaVista Translation Service page, shown in Figure 2-5.

Note *You can also get directly to this Translation Service page by clicking the Translations tab located immediately above any AltaVista Search Form. Particularly if you're translating just a few words, rather than a Web page that you found, consider using the tab to access the AltaVista Translation Service.*

The URL (of the page you clicked Translate on) appears in the text box. To translate that page and see the newly translated text in the original context (with original graphics and layout), first choose the appropriate language pair from the Translate from: drop-down list.

These are the language pairs currently available:

- English to French
- English to German
- English to Italian
- English to Portuguese
- English to Spanish
- French to English
- German to English
- Italian to English
- Spanish to English
- Portuguese to English

As other language pairs become available, they'll be added to the drop-down list.

After you select the language pair, just click Translate. Your selected page will be translated and then will appear in the browser window.

As you use this translation service, bear in mind that the translations are computer-generated—and the computer doesn't deal well with slang, bad grammar, or incorrect spelling. The best results come from news articles and journalistic pieces, but most documents using standard grammar, terminology, and spelling will be fine.

AltaVista Translation Service

SYSTRAN® TRANSLATION SOFTWARE

Make a child's wish come true this Christmas. click here WORLD CONCERN

Provide either plain text or the address (URL) of a Web page to translate:

http://www.studentenproteste.org/karte/stadt/Tuebingen.html

Translate from: German to English Translate

Help | Feedback | About

digital Digital Equipment Corporation

Disclaimer | Privacy Statement

Copyright 1997 © All Rights Reserved

Download SYSTRAN® PERSONAL

The AltaVista Translation Service translates either plain text or Web pages

FIGURE 2-5

Additionally, if you have particularly long documents to translate, the service might not translate the whole document at once. If this happens, just copy and paste sections of the document into the text box to translate the document in smaller sections.

Tip *While you're at the Translation Service page, be sure to bookmark it. You can return to it and paste in any text you'd like to translate—say, from an email message or a Usenet posting.*

Or, if you have something else in mind for a translation, like translating an email message, part of another document, or a different Web page, you can just paste the URL or complete text into the box in the Translation Service page, select the language pair, and click Translate.

SOME POINTERS TO OTHER USEFUL CHAPTERS

In this chapter, you've learned how to do most basic AltaVista searches and seen some of the other features and components that AltaVista has to offer. You'll probably find that Simple Search meets most of your searching needs. We recommend, though, that you continue on to other chapters in this book to discover all of AltaVista's power.

- Visit Chapter 3 to learn about AltaVista's People and Business Search features.
- Stop by Chapter 4 to learn how to use AltaVista's awesome Refine feature.
- See Chapter 5 to learn about doing Advanced searches.

Searching for People and Businesses

AltaVista has expanded its offerings and added new dimensions to your Internet searching—letting you search for specific people and businesses or search by category, all straight from the AltaVista site. Searching the Internet has never been easier, quicker, or more personal!

In this chapter, we take a look at these new searching features, all of which rely on their own specific specialized databases and content, rather than being merely a different way to get to the main AltaVista index. First, we'll look at AltaVista's Business Search feature, which lets you locate businesses by business type, name, or location. You can even look up toll-free numbers, which is a very handy feature when you are trying to contact businesses. Next, we'll look at People Search, which lets you search for people by name and location. (Both Business Search and People Search are available through a partnership with Switchboard.) Finally, we'll tour the Browse by Subject feature, which lets you browse for information within categories, in partnership with LookSmart. This feature offers you the convenience of a comprehensive Internet directory to supplement AltaVista's search power for your searching ease. Wow. All of this awaits you!

Note *The AltaVista team is continuously working toward making these features as easy to use as possible. Therefore, you might notice that some of the menus, buttons, or formats have changed slightly from how they're described in this chapter.*

Did We Forget Translations?

As you work your way through this chapter, you'll probably notice in the figures one of the tabs on the AltaVista Search Form—the Translations tab. And if you notice that, you'll probably also wonder why we don't cover it in this chapter; it is, after all, a prominent tab on the search form. This chapter focuses on specialized searching that does not use the main AltaVista database—business searches, people searches, and searching by category. You'll find AltaVista's Translation feature discussed fully in Chapter 2.

USING ALTAVISTA'S BUSINESS SEARCH FEATURE

It's an enormous Internet yellow pages...no, it's an enormous Internet business white pages...actually, it's *both.* Think of AltaVista's Business Search feature as being an enormous Internet business pages directory. Like your town's yellow pages, you can pick a category of business that you want to search, then look up a specific business name and location. For example, if you're in the market for car insurance, you'd look in the yellow pages under Insurance, then look for specific company names or an agent that's close to you. Or, like your town's business white pages, you can simply look up a company's name and find the information you need. That's it!

Just as your local phone book grows and changes over time, so does AltaVista's Business Search feature. The listings, which are maintained by Switchboard, Inc., are licensed from a leading compiler of business and residential listings that gathers data from published white pages directories and other publicly available sources. The data is updated every four to six months and includes thousands of business listings.

Because the Business Search (and People Search, for that matter) features draw on a specialized database, these features will likely offer you an effective supplement to regular AltaVista searches. For example, if you don't find Joe's Construction Contractors listed in the Business or People Search features, you might go ahead and search in AltaVista proper to get the contact information you need.

Note *The Business and People Search databases currently contain data only from the United States.*

Navigating the Business Search Features

Ready to get started? Great! All you have to do is click the Business Search tab, located above the AltaVista search form.

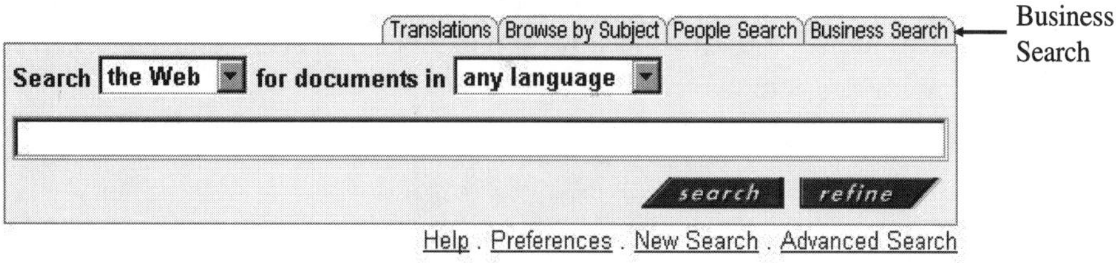

Translations | Browse by Subject | People Search | Business Search

Business Search

Search the Web for documents in any language

search refine

Help . Preferences . New Search . Advanced Search

AltaVista whisks you away to the default Business Search page, as shown in Figure 3-1.

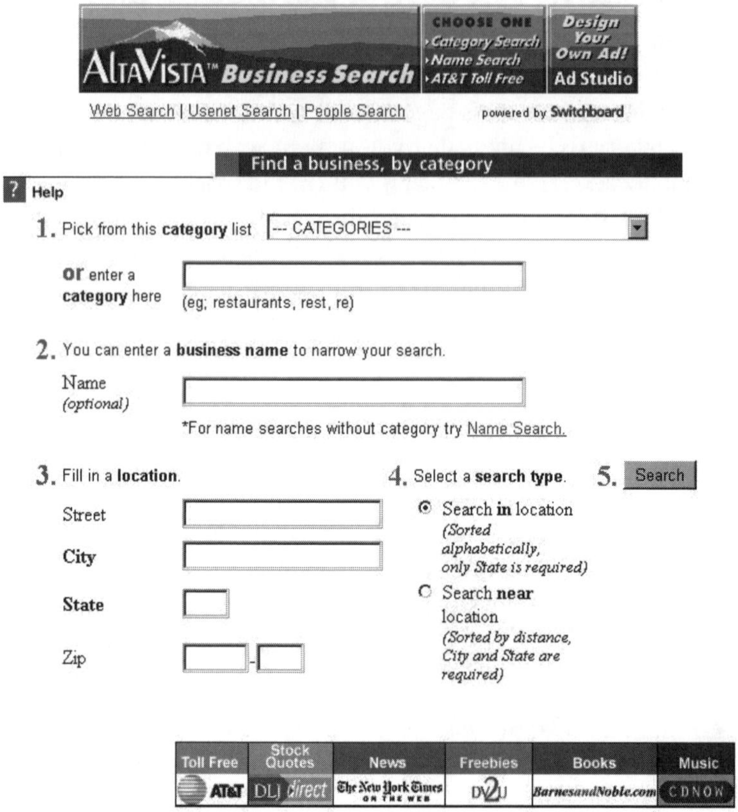

The default Business Search page

FIGURE 3-1

In this default page, you can search for businesses by category, which is a good starting point for your business searches. If you want to search by business name or search for a toll-free (1-800 or 1-888) business phone number, you can switch to those search modes by clicking the appropriate link in the AltaVista banner, as shown in the following graphic:

Searching by Category in Business Search

To search by category, make sure you've clicked the Business Search tab on AltaVista's Simple or Advanced Search pages, then follow these steps:

1. Select a business category either by choosing one from the pop-up menu or by entering a category name in the space provided.

   You might take a minute to browse through the list to become acquainted with the most popular categories listed. Most businesses are associated with a product category, such as Restaurants. The categories are quite comparable to those found in the yellow pages of a phone book.

   Alternatively, you could enter a category in the Category box by clicking your mouse in the box, then typing the category name or just a few letters of it. (Note that you must use one of the predetermined categories; you can't just create one.)

Tip *If you select a category from the drop-down list and enter a search term in the box below, the search term overrides the list selection.*

2. Optionally, enter a business name to narrow your search.

Tip *If you know the business name but not necessarily the category, start with a business name search.*

3. Type in information about the business location. The more information you can enter, the more you can narrow the search.

Note *If you enter a city, you must also enter a state. However, if you enter only a state, a city isn't required.*

4. Choose to search in or near the location you specify by clicking the appropriate radio button.
5. Click Search.

If you selected a category from the pop-up menu, you'll see a new, framed screen, with your search results in the left frame. If you happen to get a long list of search results, you can easily search through them by clicking the alphabet letters provided to view only businesses that start with that letter. In the right frame, you'll find related yellow pages-type ads, as shown in Figure 3-2.

If, back in Step 1, you opted to type in a few letters of a category name, AltaVista returns a list box of categories that contain that word or conceptually match the word's meaning. For example, if you enter *car,* AltaVista returns the Automobile as well as Carpenter categories. If you enter the first letter of a category, AltaVista returns all categories that begin with that letter, as shown in Figure 3-3.

Simply select the correct category from the menu, then click Search again.

If you want to do another Business Search, click the New Search link, located toward the top of the left frame. Or, if you're ready to return to AltaVista's Web or Usenet search, click the appropriate link at the bottom of either frame.

View a Map...Get Directions...Fast!

Did you finally find the business you've been looking for but don't know how to get there? Use the Map and Directions features, available on most search results lists. Just click the icon labeled MAP to zoom in on a map showing the location of the business you found. Or, if you need more help, click the car icon to get help with mapping a route to your business destination. By the way, these handy features are available with Search by Category searches as well as Search by Name searches.

Searching by Business Name in Business Search

If you know which business you're looking for, all you have to do is search for the business name. Here's how:

1. Make sure you've selected the Business Search tab from AltaVista's Simple or Advanced Search pages.
2. Click Name Search—either the graphic icon located in the AltaVista banner or the link under Step 2. You'll see the Find a business, by name page, as shown in Figure 3-4.

Results of a Business Search

FIGURE 3-2

3. Enter all or part of the business' name in the space provided.
4. Optionally, enter the city and state in which the business is located. Doing so helps narrow the search results.

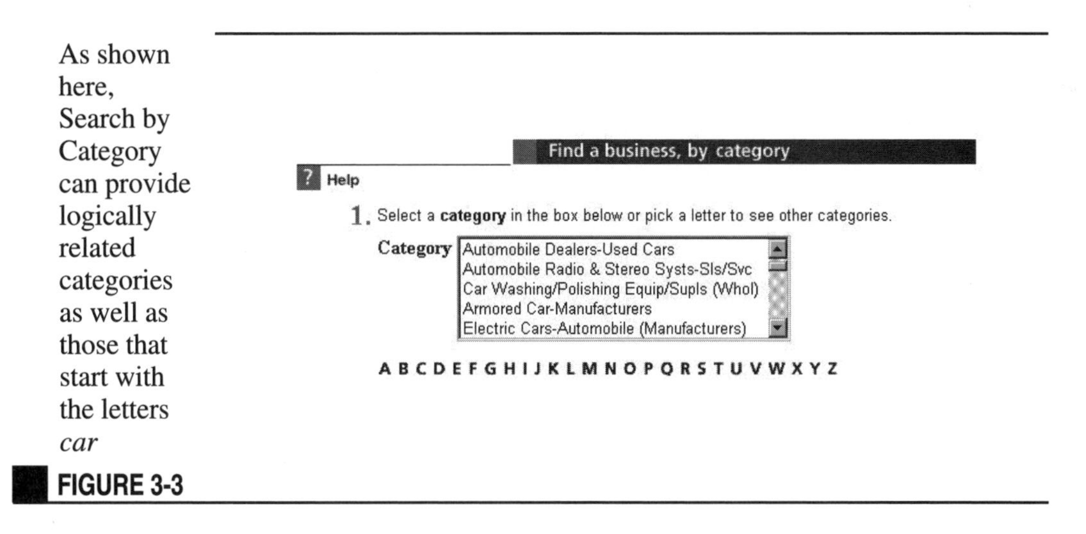

As shown here, Search by Category can provide logically related categories as well as those that start with the letters *car*

FIGURE 3-3

AltaVista™ Business Search | CHOOSE ONE › Category Search › Name Search › AT&T Toll Free | Design Your Own Ad! Ad Studio

Web Search | Usenet Search | People Search   powered by Switchboard

? Help

Find a business, by name

Enter all or part of the business name below.

Name

City   State

Search

The Find a business, by name page

FIGURE 3-4

Note *If you enter a city, you must also enter a state. However, if you enter only a state, a city isn't required.*

5. Click Search.

You'll zoom to a results page, similar to the one shown in Figure 3-5.

Listings 1 - 8 | More Listings   Modify Search | New Search

Digital 129 Parker Street, Maynard, MA 01754
Phone: null Toll-Free Number : (800)344-4825 MAP

Digital 1 Presentations 1820 San Pedro Dr NE # 11, Albuquerque, NM 87110-5956
Phone: (505)265-1013 MAP

Digital 1 TV 9 Commerce St, Williston, VT 05495-9725
Phone: (802)863-9111 MAP

Digital 2000 Communication 6004 Kissena Blvd, Flushing, NY 11355-5548
Phone: (718)886-7332 MAP

Digital 21 Inc 3100 Plainfield Rd, Dayton, OH 45432-3713
Phone: (937)254-2100 MAP

Digital 360 Inc 117 S Morgan St, Chicago, IL 60607-2618
Phone: (312)421-2668 MAP

Digital Abacus RR 1 Box 279A, Bolivar, PA 15923-9611
Phone: (412)238-7879 MAP

Digital Abacus 201 FM 2818 Rd W Trlr 37, College Station, TX 77840-6215
Phone: (409)764-5939 MAP

Listings 1 - 8 | More Listings   Modify Search | New Search

The Find a business, by name results page

FIGURE 3-5

As with other business results pages, you can scroll through the listings and choose to view a map or get directions. From here, you can choose to do a new search or modify the search you just completed by using the New Search and Modify Search links, located toward the top and bottom of each results page. Or, if you're ready to get back to AltaVista proper, simply click the link of your choice at the bottom of the results page.

Searching for Toll-Free Numbers

Using the Business Search tab, you can also find toll-free business numbers. Let's take a look:

1. Make sure you've clicked the Business Search tab from AltaVista's Simple or Advanced Search pages.
2. Click AT&T Toll Free—either the graphic icon located in the AltaVista banner or the link under Step 2. You'll see the Find a toll-free number page, as shown in Figure 3-6.

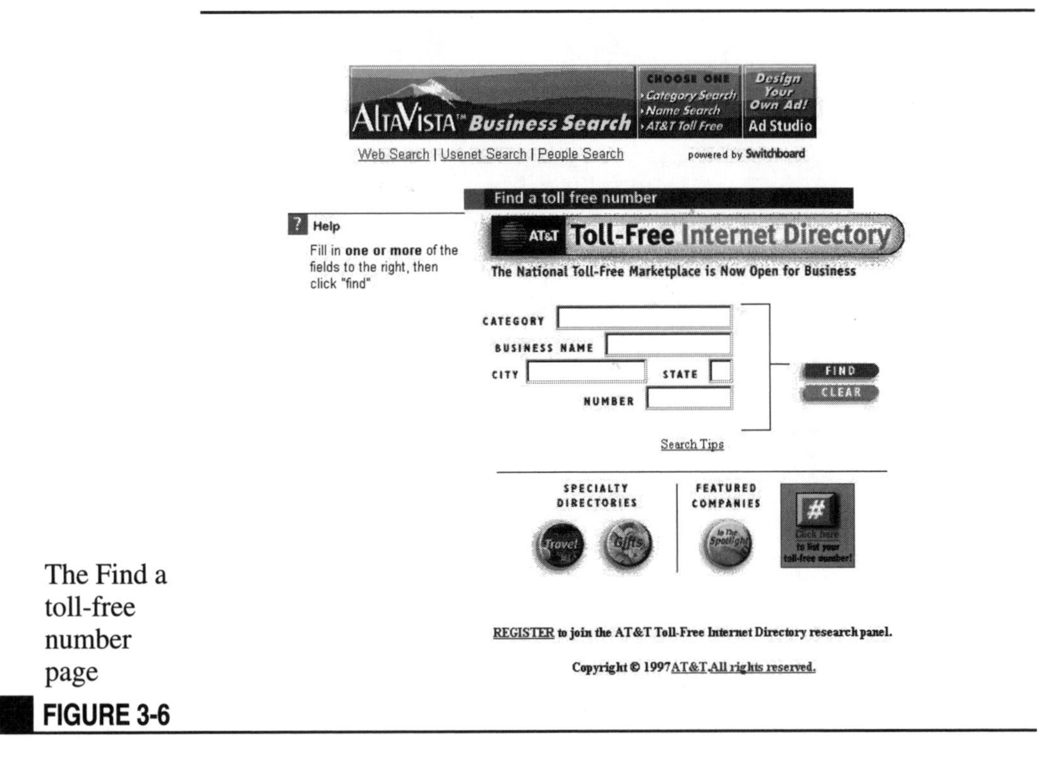

The Find a toll-free number page

FIGURE 3-6

3. Fill in one or more of the fields. The more fields you fill in, the narrower your search results will be.
4. Click Find. Tada! You'll see a results page that looks similar to Figure 3-7.

If the results list is particularly long, consider using the letters located at the bottom of the page to zoom right to the listing you need. When you're done, use the Back arrow in your browser to get you back to the AltaVista site.

USING ALTAVISTA'S PEOPLE SEARCH FEATURE

Think of AltaVista's People Search feature as being similar to your town's telephone white pages. With People Search, you can search for people by name and find information such as their phone number, address, and email address. All you have

AT&T Toll-Free Internet Directory

NEW SEARCH | INSIDE THIS SITE | ADVERTISER OPTIONS

Your Search Results

There are 7 listings for **digital:name and MA:state**. **1 through** 7 of 7 listings displayed.

Company/Directory Category	Toll-Free Number
Advanced Digital Development (Boston MA) Computers - System Designers & Consultants	800-814-4300
Advanced Digital Development Multimedia Systems (Waltham MA) Computers	800-226-7669
City Scope Digital Imaging (Burlington MA) Posters	800-447-6665
Digital Directory Assistance (Marblehead MA) Computers - Software & Services	800-284-8353
Digital Equipment Corporation (Stow MA) Computers	800-332-4636
Digital Images (Boston MA) Computers - Graphics	800-341-7878
Digital Vision (Dedham MA) Computers - Peripherals	800-346-0090

[New Search | Inside This Site | Category Index | Advanced Search | Feedback]
[Advertiser Options | Travel Directory | Gift Directory | Featured Companies]

The Find a toll-free number results page

FIGURE 3-7

to do is fill in a name and as much information as you know. AltaVista does the searching for you. Just follow these quick steps:

1. Click the People Search tab from AltaVista's Simple or Advanced Search pages. You'll see the Find a person, by name page, as shown in Figure 3-8.
2. Fill in the person's last name in the appropriate box, and fill in as much information as you can. The more fields you fill in, the narrower your search results will be.

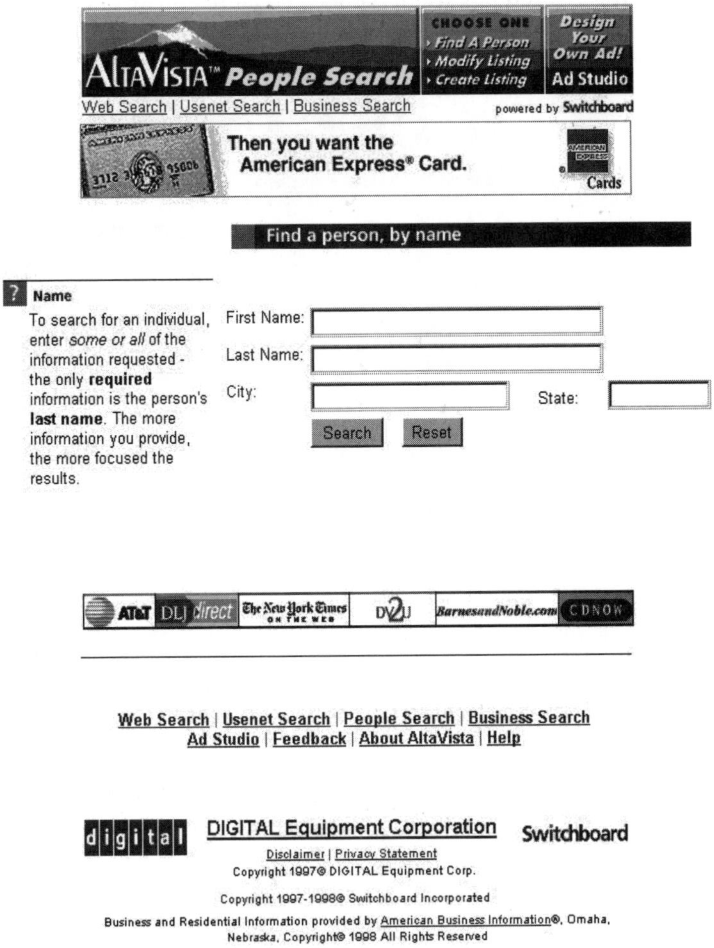

The Find a person, by name page

FIGURE 3-8

Note *If you specify a city, you must also fill in a state.*

3. Click Search. You'll see a results page that looks similar to Figure 3-9 (The addresses in this figure have been blurred to maintain privacy for the people listed.)

Pretty simple, huh? You'll notice that the results list offers you the opportunity to send the person an electronic card or even order flowers for them. Nice thought! To modify your search, use the Modify Search link, or to do another People Search, click the New Search link; both are located at the top and bottom right of the results

AltaVista™ People Search
CHOOSE ONE
› Find A Person
› Modify Listing
› Create Listing
Design Your Own Ad!
Ad Studio
Web Search | Usenet Search | Business Search
powered by Switchboard

Toll Free | Stock Quotes | News | Freebies | Books | Music
AT&T | DLJ direct | The New York Times on the Web | D2U | BarnesandNoble.com | CDNOW

ONE CLICK GETS YOU A FREE & PERMANENT E-MAIL ADDRESS FOR THE REST OF YOUR LIFE!

Listings 1 - 8 | More Listings   Modify Search | New Search

Ray, Darin L   CARD   FLOWERS

* Ray, Eric   CARD   FLOWERS

Ray, Jim   CARD   FLOWERS

Ray, Kenneth E   CARD   FLOWERS

Ray, Kristie A   CARD   FLOWERS

* Ray, Russell L

Ray, William J Jr

Raymond, Ace S   CARD   FLOWERS

*Denotes Switchboard member

Listings 1 - 8 | More Listings   Modify Search | New Search

The Find a person, by name results page

FIGURE 3-9

page. When you're done with People Search, click the link of your choice at the bottom of the search or results pages.

Note *The databases used for the People Search feature rely on published white pages across the United States; results might not reflect recent changes of address or people with unlisted numbers.*

USING ALTAVISTA'S BROWSE BY SUBJECT FEATURE

For a completely different take on finding information on the Web, try out AltaVista's Browse by Subject feature. Browse by Subject lets you scan through a wide selection of categories of information from the Web—much like browsing through a bookstore by looking for the little signs that tell you what kind of books to expect on each shelf.

As you read through this section, keep in mind that the Browse by Subject feature offers you a collection of selected Web resources that were chosen for their quality and usefulness. Browse by Subject does not, however, show you all of the resources available on the Web. For that, you'll need to search carefully through AltaVista proper.

So, you're asking, if Browse by Subject doesn't let me see all of the pages on the Web, why would I want to waste my time with it? To get a broad view of the information available on the Web or to find some useful resources as a jumping-off point. For example, suppose a young child—yours or someone else's—is starting to show an interest in computers and wants to play on your computer whenever the opportunity arises. You want to encourage this, but need more information about educational software for kids. An AltaVista Simple Search for *kid* child* education* software* yields over 750,000 matches, with no real context for what's useful or not, and very few substantive-looking sites. That, in a nutshell, is the problem with full-text search of the Internet—it's great if you're looking for something specific, but if you're just browsing, you could be busy for quite some time.

What's the alternative? A smaller—selective—collection of Web sites, organized by category. That's Browse by Subject! You can look for a category with *kids educational software* in it, then check out the sites available. You might find exactly what you were looking for, or you might just get some additional information that would let you return to AltaVista and try a much more focused, targeted search.

So, where did Browse by Subject come from? It's from the LookSmart directory, actually. LookSmart is a guide to the Web, with a specially selected list of Web sites. Editors hand-select sites for inclusion in the directory, categorize them appropriately, and provide mini-reviews. AltaVista licensed the technology and use of the directory

to help provide AltaVista users with a comprehensive searching solution. Particularly with broad topics, browsing through the editors' picks can help provide some useful perspective.

How do you get started using Browse by Subject? Read on!

Using Browse by Subject

Now that you're ready to take a browse around, just follow these easy steps for using the Browse by Subject feature:

1. Click the Browse by Subject tab from AltaVista's Simple or Advanced Search pages. You'll see the Browse by Subject page, as shown in Figure 3-10.
2. Select a subject from the high-level subject categories at the left of your screen. AltaVista will present a selection of sub-topics.

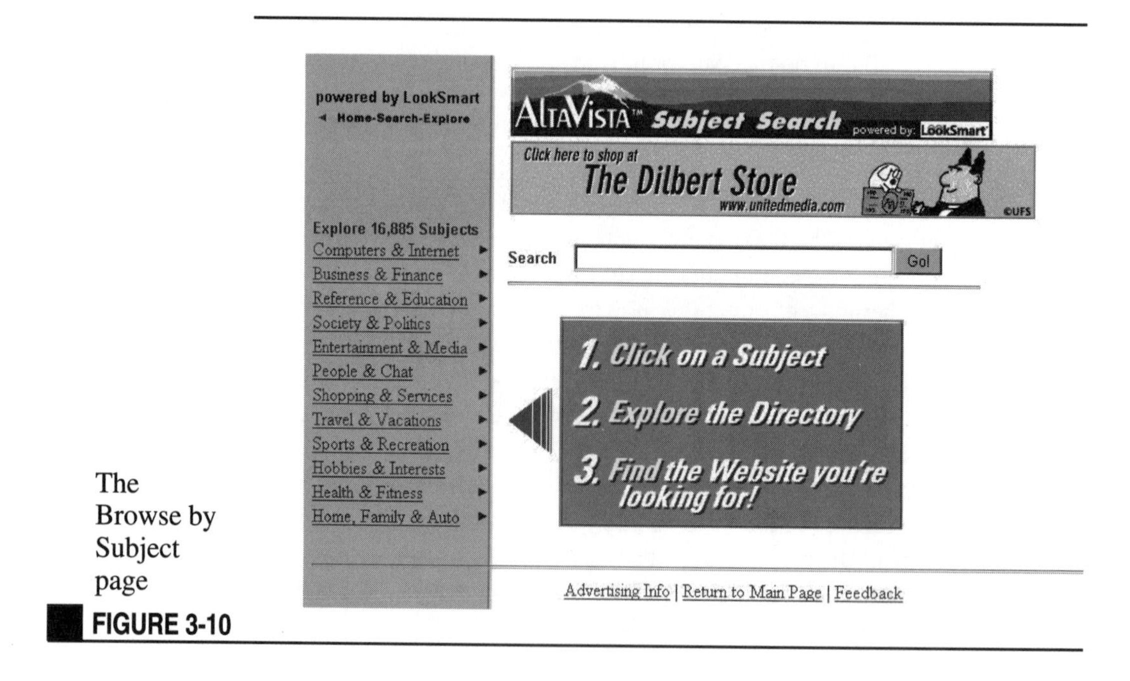

The Browse by Subject page

FIGURE 3-10

3. Continue to burrow through sub-topics until you get to the information you need. Subtopics lead to sub-topics, which lead, eventually, to actual information.

Note *The small arrow icon indicates that another submenu awaits. A small white box with horizontal lines represents documents, and shows where you can expect real results.*

Now, that's easy! If you're more comfortable doing searches than you are poking through menus and submenus, try using the Search field. This field lets you type in search terms for the Browse by Subject database. If you type in something that's in the Browse by Subject database, you'll see results only from that database; however, if you search for something that's not in the database, you'll get results directly back from AltaVista.

Now that the pieces have all fallen into place, let's try it out. Start by looking for the information mentioned previously about children's educational software. You could either search or browse. Start by browsing—probably within Computers & Internet.

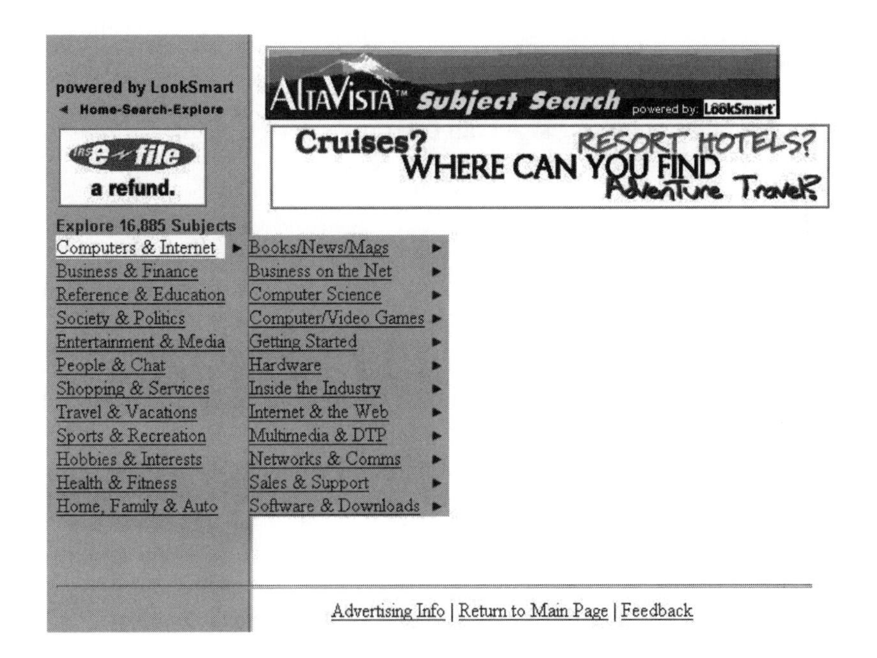

From Computers & Internet, Software & Downloads looks like a logical next step. Clearly, Kids is the way to go from here.

And, finally, the Education link ends with a little document icon (and not another arrow), so it's our answer.

Look at all those choices, and all carefully selected by the LookSmart staff, as shown in Figure 3-11.

Note *You can click on any of the categories at the top of the matches to backtrack to a specific category or subcategory from the journey.*

Alternatively—if you're not into browsing right now—you could just search. Click the Powered by LookSmart link to return to the main screen if you want, or just search in the Search field at the top of the window.

To find the same stuff—kids educational software—you could just type in *kids educational software.* You'll see a list of matches, similar to the results list above, with a page header telling you how many matches are shown and that they're from LookSmart.

A search like this, as shown in Figure 3-12, gives results from the Search by Subject guide first, then lists Web sites from the AltaVista index. If no sites from the Search by Subject guide match the query, then only sites from the AltaVista list appear.

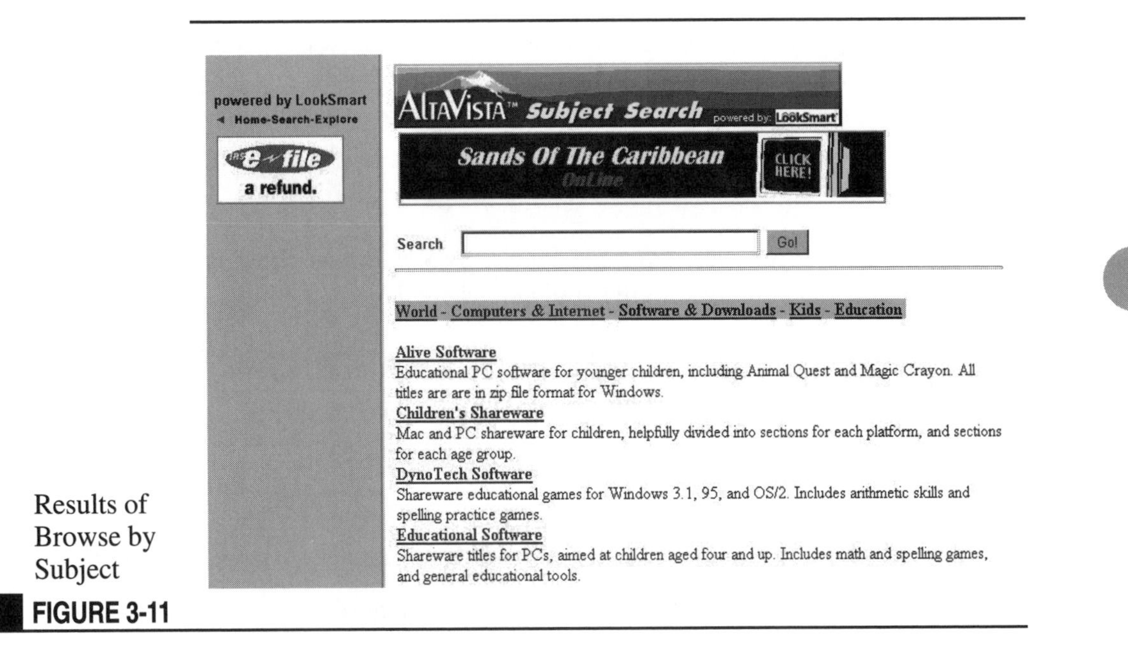

Results of Browse by Subject

FIGURE 3-11

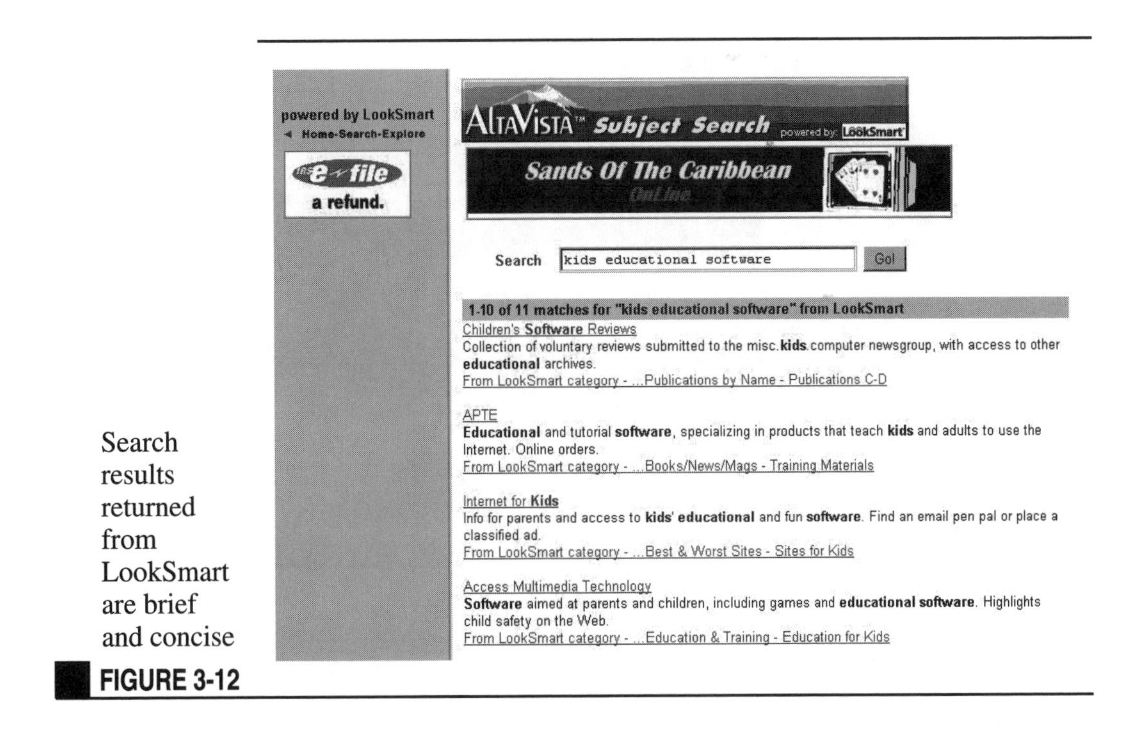

Search results returned from LookSmart are brief and concise

FIGURE 3-12

The only potential problem with searching within Browse by Subject is that a seemingly minor change in search terms can cause lots of problems with your results. For example, if you search for *kid educational software—kid* singular rather than plural—you'll get matches from the entire Web, just as you'd get from AltaVista proper. See Figure 3-13 for an example.

Remember *If you search through Browse by Subject, watch the line at the top telling you how many results were found and where they came from.*

SOME POINTERS TO OTHER USEFUL CHAPTERS

In this chapter, you've learned how to do specialized searches at AltaVista, including business searches, people searches, and searches by category. So, where should you browse next?

- Stop by Chapter 4 to learn how to use AltaVista's awesome Refine feature.
- Visit Chapter 5 to learn about AltaVista's Advanced Search.
- See Chapter 6 to learn about doing Simple and Advanced Usenet searches.
- Peruse Chapter 9 for sample searches and nifty search ideas.

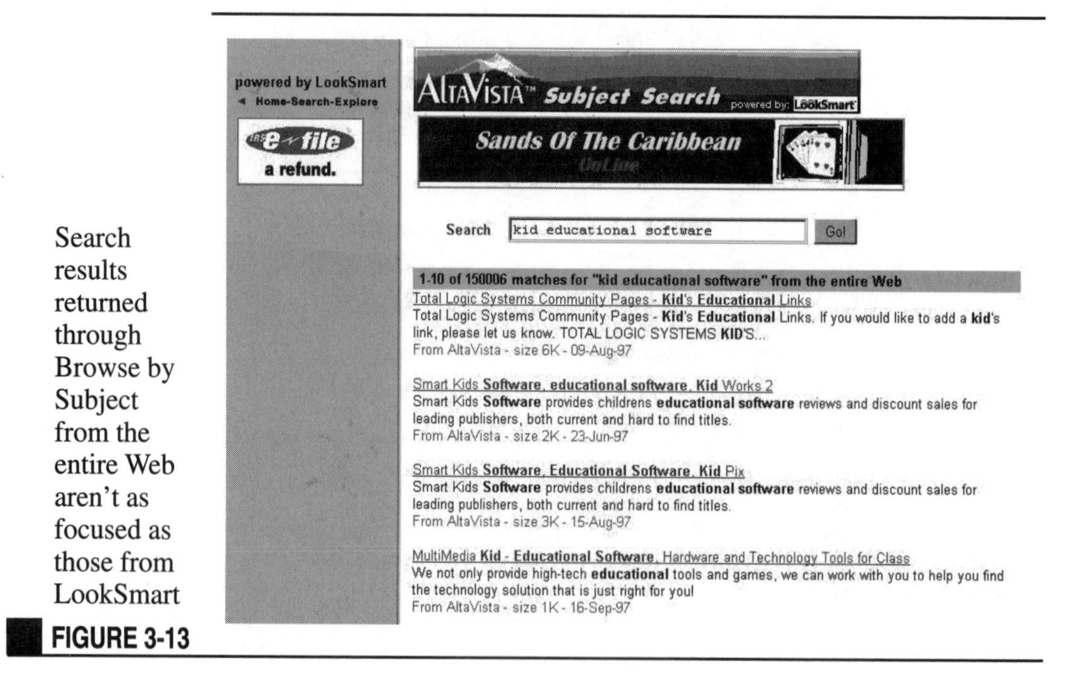

Search results returned through Browse by Subject from the entire Web aren't as focused as those from LookSmart

FIGURE 3-13

Refining Your Searches

It started as "Cow9." The code name came likely more through sleep deprivation than anything else, but the expression "on Cloud Nine" (for extreme happiness) certainly played a role. After all, this technology can revolutionize searching on the Internet once again. Of course, the test searches on **BSE** (a.k.a. Mad Cow Disease) to prove the value of the technology solidified the name. Now, though, you might know this AltaVista feature as Refine.

Suppose you've just run a Simple or Advanced search of the Web. Scanning through the first few screens of results, you see that there seem to be several good matches. But there also seem to be lots of matches that don't meet your needs. While you could refocus your search—by adding or excluding words or grouping words together, in either Simple or Advanced Search—AltaVista now offers a better way: Refine. AltaVista's unique search technology identifies, then groups, the major topics and subtopics that appear in the results list, and allows you to pick which of the topics you want to see—and which ones you don't. Refine offers a different and quite effective way to narrow the scope of results to just the items you want, removing ambiguity from searches, and to help find related terms that might also help the search.

In this chapter, we'll show you how to get started with Refine, teach you how to use its List and Graph views, and tell you a bit about how Refine works. Never before has narrowing your Web searches been easier—or more effective. Let's get started!

Note *Refine is available for Web searches only.*

GETTING STARTED WITH REFINE

Think of Refine as a search assistant that groups search results into related topics to make the information easier for you to use and manage. It does all the grunt work for you. It reads your search results in a fraction of a second and then sorts the results into customized topics, each of which organizes sets of related documents.

Although you could—and usually should—also narrow your searches by adding, deleting, or combining search terms, using Refine makes the task even easier by working on the broad topics themselves, rather than on individual terms. All you have to do is review the topics, then select the items or topics that meet your needs. AltaVista then gives you a new results list, which includes only the topics you specified.

With just a few easy steps, you can narrow your searches to just the information you need. Here's the basic process:

1. Do a search (or just enter the search terms in the Query field).
2. Click Refine.
3. Choose topics to include or exclude.
4. Click Search or click Refine again.
5. Repeat steps 3 and 4 as needed.

Tip *Refine works much better with a broad search that returns lots of results than with a more focused search. The statistical analysis that supports Refine doesn't work nearly as effectively with small numbers of matches. Of course, if you only get a few matches, you could just look at the results—because they probably don't need to be refined.*

Let's take a look at a sample search. Suppose you do a Simple Search for **cat**. As shown in Figure 4-1, the results list includes everything from movie reviews to cat psychology to an airplane history.

You might have been looking for specific companies with *CAT* in the name, the UNIX command cat, CAT scanners, or information about pets, but you were not likely looking for all four. With either Simple or Advanced Search, you narrow the search by scanning through the results and manually including or excluding items. With a search results list like this, though, it's hard to tell where to even start narrowing it down.

That's where Refine comes in. Using either Refine's List or Graph views, you can narrow the results to just the information you need. Just click the Refine button, as shown here:

Translations | Browse by Subject | People Search | Business Search

Search the Web for documents in English

cat

search refine ← Refine

Help . Preferences . New Search . Advanced Search

Narrowing Searches in List View

When you click Refine, AltaVista displays the refined topics in List view, which shows the list of topics and sample words from each. As you can see in Figure 4-2, Refine does a great job of grouping the feline-related items together and the CAT scan-related items together and the UNIX command items together, and throws in a few topics that seem to have little bearing on the discussion at hand.

1189372 documents match your query.

Click to find related books at **amazon.com**

71. The Burmese Cat
The Burmese Cat. Origin. There is a legend that this breed comes from Burman monasteries where it was venerated as a divinity. A book of poems from the...
http://www.hss.caltech.edu/~eloisa/burmese.html - size 3K - 3-Aug-95 - English - Translate

72. E! Online - Movie Review - That Darn Cat
nbsp; • First Look • The Dotted Line • E! Files. ...
http://www.eonline.com/Reviews/Movies/Video/Leaves/0,79,192,00.html - size 11K - 17-Aug-97 - English - Translate

73. Clan of the Cave Cat Ring
Clan of the Cave Cat Ring. Cats have long been considered magickal creatures. Join our web club and ring for magickal cats and their people. Appropriate...
http://www.geocities.com/~zlioness/cavering.html - size 6K - 9-Aug-97 - English - Translate

74. Cat pictures
Cat picture gallery. [Pictures page 1][Pictures page 3] Bedrock's Beauty Bell A beautiful sealpoint Birman female from Bedrock's cattery, owned by Bitte...
http://www.algonet.se/~sillen/bild_en2.htm - size 2K - 29-Jun-97 - English - Translate

75. Cat Psychology
A Brief Introduction to the Psychology of Cats. You know something? In all my years of being around people, I have had the remarkably non-unique pleasure..
http://falcon.cc.ukans.edu/~malone/origin/cats.html - size 6K - 2-Mar-97 - English - Translate
http://falcon.cc.ukans.edu/~malone/origin/cats.html - size 6K - 2-Mar-97 - English - Translate

76. THE HILLBILLY CAT & ME
THESE AREN'T FROM MY COLLECTION, BUT I WISH THEY WERE!LOTS MORE TO COME! ELVIS DOLLAR BILL 197? ELVIS 3D MAGNET. ELVIS GUITAR CD RACK. ELVIS SLIPPERS....
http://home1.gte.net/eap/cool.htm - size 1K - 23-Jul-97 - English - Translate

77. Thundoer Cat
Thundoer Kat. Name: Thundoer Yang B-boy Name: TK "Thundoer Kat" Comments: one of our best b-boy... ghas lots of flavor!! Quote: "Thas hella...
http://members.aol.com/khaoscru/tc.html - size 3K - 7-Aug-97 - English - Translate

78. Ernie, The Incredible Flying Cat
Ernie, The Incredible Flying Cat.
http://www.duke.edu/~mt/ernie.htm - size 222 bytes - 10-Nov-96 - English - Translate

79. Stichting Cat Air
Stichting Cat Air®. Experience The Difference! History about the PBY-5A from Cat Air.(Dutch) History about the PBY-5A #2459 writen by Ragnar J....
http://www.worldaccess.nl/~boelpvt/cat-home.htm - size 2K - 29-Jul-97 - English - Translate

Searches sometimes produce results too varied to be usable

FIGURE 4-1

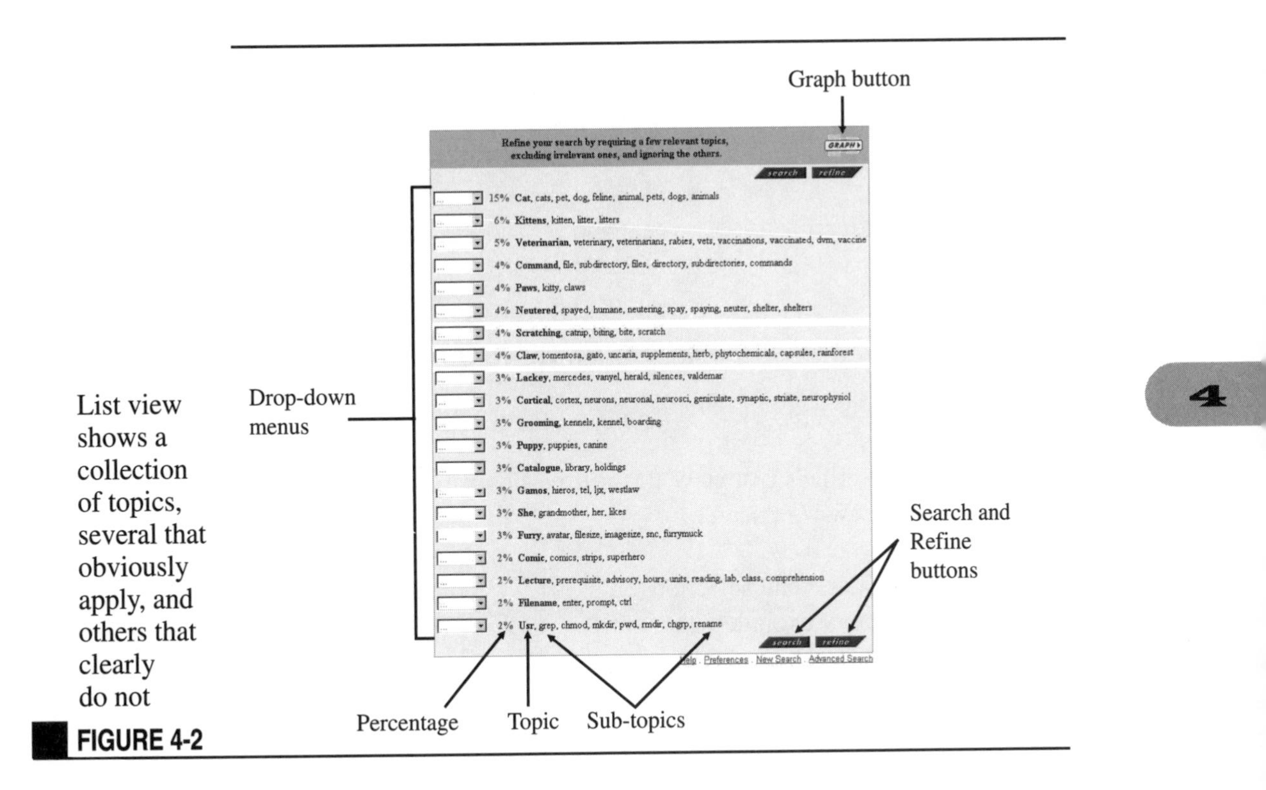

List view shows a collection of topics, several that obviously apply, and others that clearly do not

FIGURE 4-2

Note *The statistical analysis process used to generate topics sometimes finds patterns that aren't obvious or particularly useful—especially if the query does not include enough topics for a statistically significant analysis. Just ignore any real oddities in Refine lists, or try to broaden your search if you're finding more anomalies than useful results.*

Figure 4-2 also shows the Refine List view features, and the following table describes what each part of the Refine page does:

Refine Feature	Description
Graph button	Takes you to Refine's Graph view. We'll discuss Graph view in detail later in this chapter.
Drop-down menus	Lets you require or exclude topics. The default setting of ... indicates no preference.
Percentage	Indicates what percent of all the documents focus on a particular topic.

Refine Feature	Description
Topic	Shows the general fields of information resulting from the search query. This is the term in bold type.
Sub-topics	Further describes what information you'll find within general topics. These are the words in normal type.
Search and Refine buttons	Lets you submit a query based on Refine choices you made or refine the search results again.

The key to Refine's List view are the drop-down menus, which give you choices to help you narrow your search:

- The default menu selection, ellipses (...), indicates no preference—either to require or exclude—the topic.
- *Require* specifies that the topic be included in the refined search results.
- *Exclude* specifies that the topic be eliminated from the refined search results.

Note *The list of categories you see are presented (approximately) in decreasing order of prevalence in your search.*

You do not need to make a selection for each topic. If you don't specifically want to require or exclude a topic, leave the ... selection showing in the drop-down menu. The best approach is to make the most obvious changes—that is, require or exclude topics you know for sure should or should not appear in the refined results. For example, from this search, you might require a couple of clearly feline topics, and exclude a couple of the most obviously irrelevant topics. Then, click Refine again, and you'll find a much narrower results list, as shown in Figure 4-3.

Note *Refine's Require and Exclude commands affect topics, not individual items. That is, even if you choose to exclude a topic, you may still see occasional items from that topic appear. Likewise, if you require a topic, some individual matches might appear relevant to the topic yet not include the specific term you were seeking.*

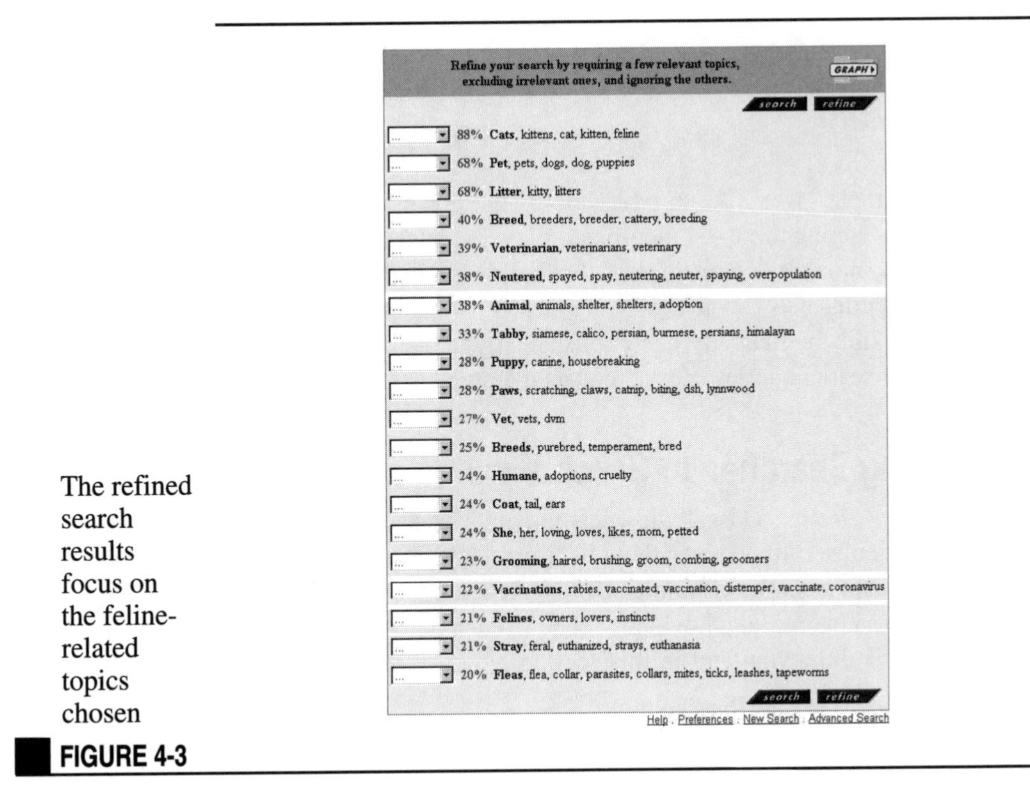

The refined search results focus on the feline-related topics chosen

FIGURE 4-3

From here, you can either continue using Refine to narrow the search, or click Search to submit a new search, enhanced with the settings from your passage through Refine. To complete this example, clicking Search returns you to a results screen, as shown next, with much more focused results. Within the search form, immediately below the query terms field, you'll see an additional checkbox, which indicates that you refined (modified) the query—in this case, by including cat…, kittens…, and excluding: command…, cortical…, and usr…

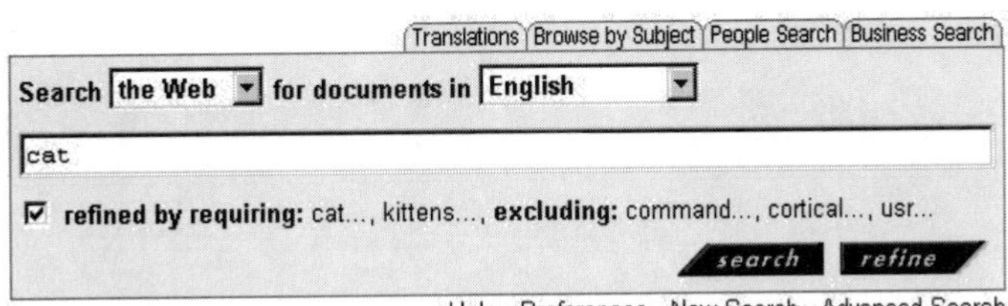

Tip *Refining a search may be an iterative process, in which you use Refine, then Search, then Refine again—depending on what you seek, the iterations could continue.*

This example gives you an overview of the power and flexibility of Refine. Rather than slogging through numerous irrelevant matches to your Web searches—like UNIX commands and businesses that have *CAT* in their names, but nothing to do with potential pets—you now have nearly exclusively appropriate matches. While you could, theoretically, accomplish nearly the same thing with trial and error and complex search queries, Refine makes it a quick and easy process.

Narrowing Searches in Graph View

Like List view, Refine's Graph view lets you narrow search results with just a few clicks of the mouse. Graph view, though, offers a few other cool features that you're likely to find handy. For example, Graph view, as the name indicates, lets you visualize the categories and how they relate to each other (as shown in Figure 4-4). Also, as we'll show you later in this section, you can require or exclude specific

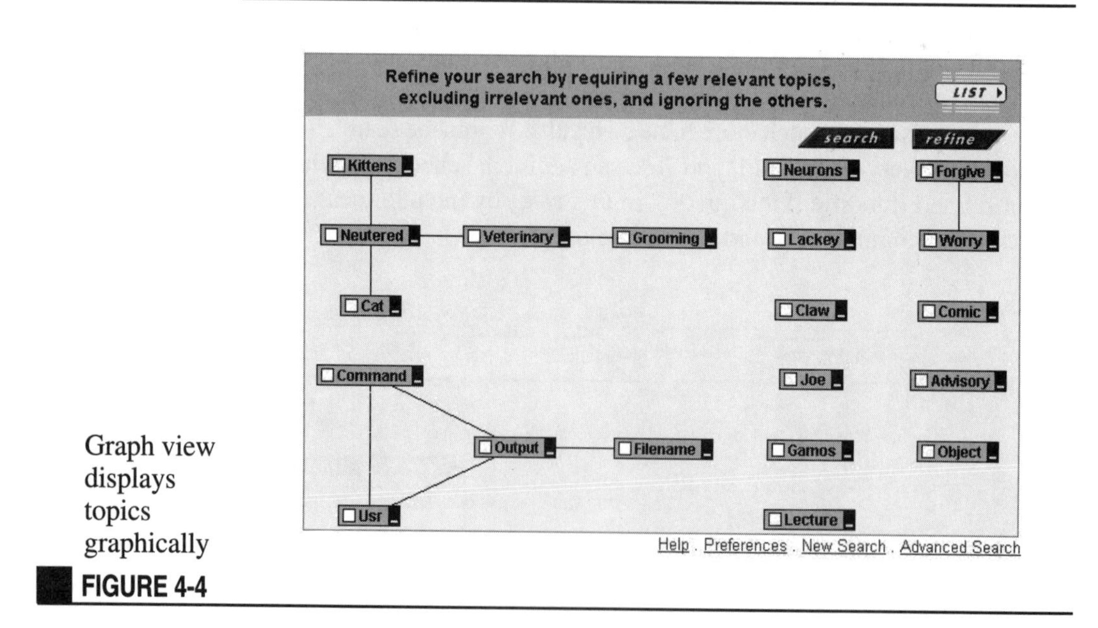

Graph view displays topics graphically

FIGURE 4-4

subcategories of information, which helps narrow searches to precisely the information you need.

To use Refine's Graph view, click the Graph button at the top right of the Refine List view screen. You see a figure similar to Figure 4-4. Note that the Graph view topics shown in Figure 4-4 correspond to the List view topics shown in Figure 4-2.

Tip *You can return to the List view by clicking the List button at the top-left corner of the Graph view screen.*

Just as with List view, all you have to do is specify which topics you want to require or exclude—or leave as is. To help you determine how relevant the topics are to your search query, check out the number of yellow bars to the right of each topic name. The more yellow bars, the more likely the topic matches your search needs. The yellow bars correspond roughly to the percentage listed in List view.

You can also determine which topics best match your search by hovering your mouse over a topic. You'll see a list of sub-topics, which gives you details about what information the topic includes.

To the left of each item (both the main topics and the sub-topics) is a checkbox, which is what you use to require or exclude the items. Click the checkbox next to a topic once to include all its sub-topics. Click it twice to exclude all the sub-topics. Click it three times (or not at all) to return to the default setting, which doesn't specifically require or exclude the sub-topics.

You can also include or exclude particular sub-topics. Simply hover over a topic to expand the list, then click once to include the sub-topic, twice to exclude it, and thrice to return to the default setting, as described previously.

- Click once to require a topic or sub-topic.

Kittens

- Click twice to exclude a topic or sub-topic.

Neurons

- Click three times to return to the default selection, which doesn't specifically require or exclude the topic or sub-topic.

Veterinary

If you have selected all the sub-topics, a green check appears next to the main topic. A red X appears if you have excluded them all. If you have a mixture of selections and exclusions, you'll see a black box within the topic checkbox, as in Figure 4-5. This indicates that you made some selections within the topic, but did not either require or exclude the topic as a whole.

Note that the Graph view uses lines to connect related topics. On some searches, you might find several disconnected clusters, while others, like this example, focus clearly on a couple of sets of topics.

Tip *If the topics seem crammed together, just click and drag on the topics to spread them out a bit.*

Although List view is the default Refine view, some people choose to make Graph view their default setting. You can easily do so—just go to Preferences from any AltaVista screen, set the preferences to Graph view in Refine, and bookmark the resulting page. See Chapter 7 for more about setting AltaVista preferences.

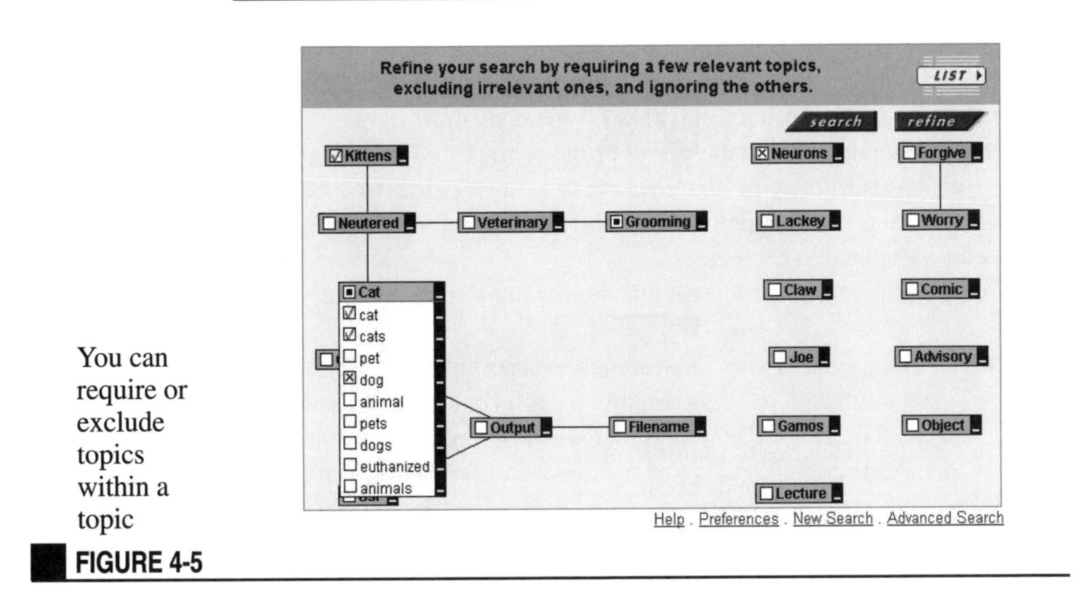

You can require or exclude topics within a topic

FIGURE 4-5

UNDERSTANDING HOW REFINE WORKS

The Refine tool was developed collaboratively by researchers at Digital Equipment Corporation and Ecole des Mines de Paris. Refine uses statistical analysis—not human interpretation—to sort the results from your searches into general topics. It analyzes the contents of documents that meet your original search criteria and displays groups of additional words, culled from the matches to your search. These topics are dynamically generated from words that occur frequently in the documents that match your initial search criteria. Topics appear in order of relevance, and words inside a topic are given in order of frequency of occurrence.

This dynamic generation of related topics, tailored to each search, distinguishes Refine from other Web search aids, which offer predetermined topics or structures into which queries must fit. As the Web and its contents change, the topics and search results AltaVista's Refine provides change automatically, making AltaVista's usefulness and timeliness unmatched.

Using Refine is similar to requiring or excluding terms in Simple and Advanced searches. You specify which words to require or exclude, and AltaVista returns results with or without those terms. The results you get with Refine, however, are broader than those you get using operators such as +, -, AND, or NOT. With search operators, an excluded term will not appear in your matches list, whereas an included

item must appear. With Refine, you require or exclude entire categories of information—not just a specific word or phrase.

Because of this, Refine results tend to be broader than results generated using operators. If a specific item on the Web matches most of the search and require criteria, yet nonetheless contains one of the terms listed under an Exclude topic, it may yet appear in your matches list, your choice to exclude that topic notwithstanding. This happens infrequently, and usually when the overall topics are very closely related to each other.

Some of the most useful techniques for using Refine include:

- Getting "good" terms for searches, as a sort of online searching thesaurus. For example, if you're searching for information in an area you don't know much about, you might not know the precise terminology or phrases used in formal papers. However, if you do a search and refine with the broader or more common terms, the formal or preferred terminology will almost certainly show up in the Refine results, thus providing the additional terms and information you need for the search.
- Finding the central papers in a field, to get a starting point for academic research. Refining "gently" into a scientific domain and carefully focusing on a specific idea will eventually yield the papers or documents with most of the good search terms—likely reference-based research papers.

SOME POINTERS TO OTHER USEFUL CHAPTERS

In this chapter, you've learned how to use Refine, AltaVista's hot new search technology. Using Refine, you can visualize which topics result from your searches and easily narrow results to just the ones you need. Where to now?

- See Chapter 5 to learn about doing Advanced searches.
- See Chapter 6 to find out how to do Simple and Advanced Usenet searches.
- Visit Chapter 7 to learn how to customize AltaVista.
- Breeze through Chapter 9 for search ideas and samples.

Advanced Search

Despite the name, Advanced Search isn't hard to use. In fact, it's almost as easy to use as Simple Search. You'll find that a lot—nearly all, actually—of the techniques you use in Simple Search also apply to Advanced Search. "Advanced" just means that you get somewhat more control over your results as the payoff for a little extra setup effort.

Why did the AltaVista Search designers provide two different ways to do something very similar? Because people's searching styles and preferences differ. Simple Search requires less thought and organization to create a quick yet effective search. Advanced Search requires more structure; it takes a little more effort and time to construct a good search, but you get a little more control over the results. That said, even the AltaVista developers normally use Simple Search, unless they need to restrict the search based on dates or to precisely control the ranking of the matches, which are both features exclusively found in Advanced Search.

Remember *Unless you particularly like using Boolean searches and are very familiar with them, you should probably stick with Simple Search, which provides all of the power but with less effort on your part.*

The key difference between Simple Search and Advanced Search is the query: Simple Search accepts queries in normal, conversational language, then it interprets your search, ranks the results, and returns the most likely matches. Advanced Search requires a more precise query in which you specify the relationships among the search terms as well as the ranking and organization of the results. Both Simple Search and Advanced Search examine the same data and find the same sites—the difference is that the Advanced Search results reflect the ranking and order you specify, rather than AltaVista's best guess about what you want.

Let's take a look at an example. Suppose you're preparing for a vegetarian potluck dinner; you might inventory the refrigerator and decide to search AltaVista for all recipes that include zucchini, corn, artichokes, and asparagus. You would probably be looking for a recipe with any one (or several) of those ingredients. You could do a Simple Search using a natural-sounding query such as *recipe with zucchini, corn, artichokes, and asparagus*—just the way you thought about it. Simple Search treats words like *and* or *or* as common English words, not commands.

You may get matches that contain any combination of the words *recipe, zucchini, corn, artichokes,* and *asparagus.*

As you'll see in this chapter, however, Advanced Search interprets queries more literally. The *and* in an Advanced Search is a command (called an *operator*) that tells AltaVista to require the word in search results. So, *and asparagus* in a query would specify that the word *asparagus* must appear in all the search results, just as *+asparagus* would require the word in all results of a Simple Search.

This is just one example of how Simple and Advanced Search differ. In this chapter, we'll show you other ways that the two searches differ and help you determine when Advanced Search is right for your search needs. Also, we'll introduce you to the Advanced Search screen, introduce you to the Advanced Search operators, and show you how to develop different types of Advanced Search queries. As always, you'll find plenty of sample searches to model your own after.

Note *You don't have to master Simple Search before you tackle Advanced Search. We assume, though, that you are at least familiar with Simple Search; some explanations in this chapter clarify Advanced Search in terms of how it differs from Simple Search. If you're new to Simple Search, you might refer back to Chapter 2 as you're browsing through this chapter.*

WHY USE ADVANCED SEARCH?

If you worked your way through Chapter 2, you probably found that Simple Search is easy to use and provides good results. Simply entering a natural-language query—or even a question—gives you useful matches. And, including plus (+) and minus (-) signs is an easy way to require or exclude terms from the results list. In fact, most AltaVista searches use only two to three words in Simple Search.

So, why use Advanced Search? Advanced Search, which is almost as easy to use as Simple Search, provides additional capabilities, such as precisely narrowing your search or searching by date. For example, you can use Advanced Search to do any of the following:

- Search for a word or phrase that occurs *near* (not necessarily immediately beside) another one
- Organize long, complex queries
- Restrict your search based on dates

- Retrieve more than 200 matching items
- Count items on the Web

If you have any of these search needs, then Advanced Search is best. The only drawbacks are that you'll need to spend a bit more time developing the search query than you would with Simple Search, you'll have to become familiar with the Advanced Search page, and you'll have to learn about operators. We'll explain these topics in the rest of this chapter.

ABOUT THE ADVANCED SEARCH PAGE

Let's start by looking at the AltaVista Advanced Search page to help familiarize you with its features. Launch your browser and connect to AltaVista (type in **http://altavista.digital.com**). You'll see the same Simple Search page that you used in Chapter 2. Click the Advanced Search link, immediately below the Simple Search form, to go to the Advanced Search screen. The link to select is indicated here.

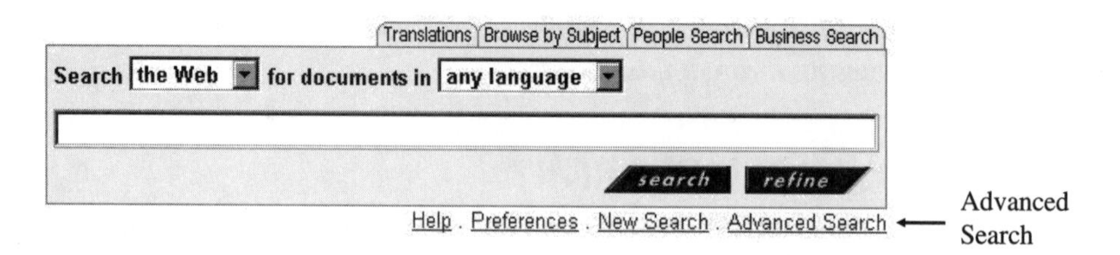

When the Advanced Search form comes up, you'll see something like the screen shown in Figure 5-1.

Tip *Bookmark the Advanced Search page so you can quickly access it time after time!*

The AltaVista Search Navigation Bar at the top of the screen lets you jump to Simple Search via the AltaVista Search Network with a click on the logo graphic. There are also a couple of ever-changing promotions for other sites that you can link to. (This navigation bar is identical to the one you see when you are in Simple

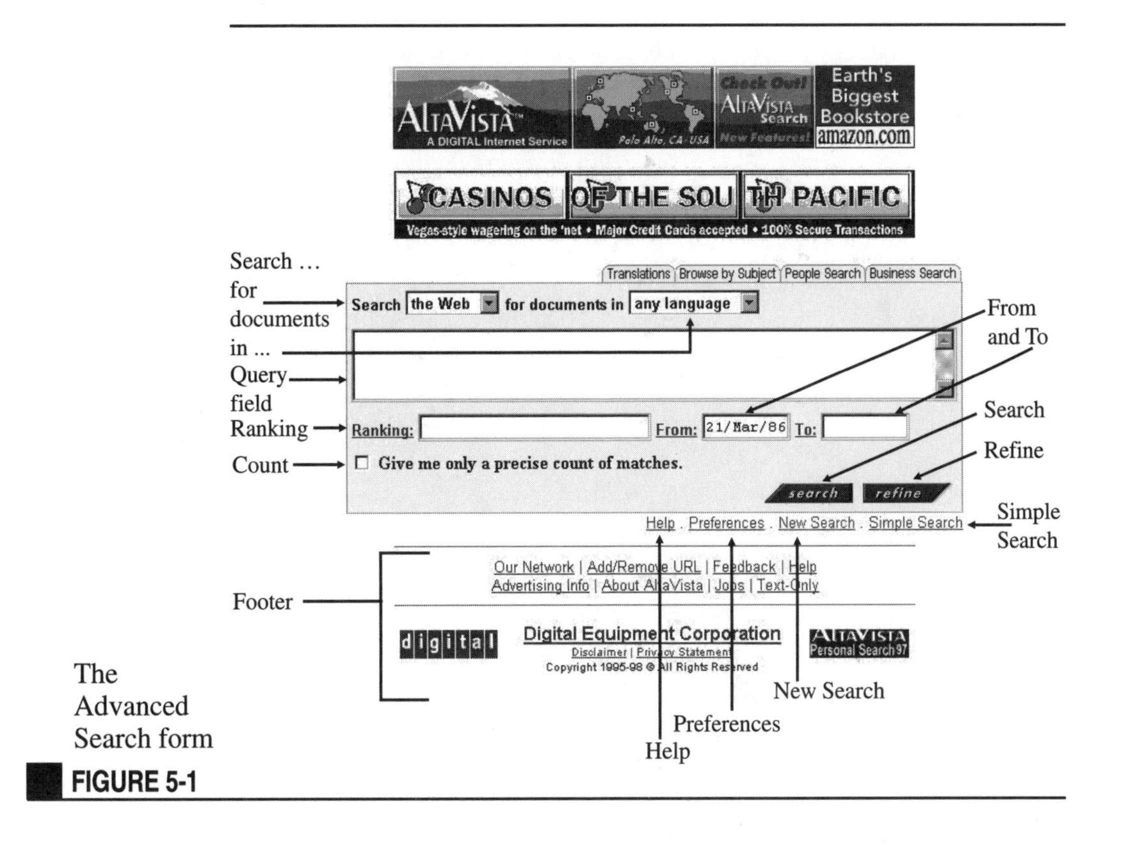

The Advanced Search form

FIGURE 5-1

Search.) The remainder of the screen is slightly different from Simple Search. The following list describes the fields and links available in Advanced Search.

Search … for documents in …	Works the same in Advanced Search as in Simple Search. You choose if you want to search the Web or Usenet as well as selecting the language you want to get results in.
Query field	Lets you enter the search expression (including the term or terms for which you want to search and, optionally, any operators you require).

Ranking	Lets you specify the terms that should be ranked highest (presented first) in your results list.
From and To	Lets you enter optional starting and ending limits for the documents returned in your results.
Give me only a precise count of matches	Lets you request only an accurate total, and not the actual matches. (The results count is explained in more detail later in this chapter.)
Search	Lets you begin the search. In the Simple Search, you could submit the search by pressing ENTER (in most browsers), but in Advanced Search, you must click on the Search button.
Refine	Lets you focus your search results more precisely—it works just as it works with Simple Search. See Chapter 4 for the specifics.
Help	Links directly to the AltaVista Advanced Search Help files. If you're referring to this book, you're unlikely to need the help; however, you'll find updated information and ideas that can get you over minor hurdles.
Preferences	Lets you set AltaVista preferences to affect future searches. The Preferences are the same as the options available from the Simple Search screen.
New Search	Clears your current search and presents you with the same blank screen you viewed when you first selected Advanced Search.
Simple Search	Returns you to the Simple Search screen.
Footer	Like the Simple Search footer, this provides links to pages where you can send feedback, add a URL, get a Text-Only search form, and other options.

Note *The results you get from Advanced Searches will look just like those from Simple Searches. Refer to Chapter 2 for more information about the results format.*

About the Results Count

When AltaVista returns a list of search results (from a Simple or Advanced Search), it also tells you how many times the query words appeared and in how many documents. That count is only an estimate, but sometimes reporters, graduate students, librarians, and trivia fans try to read more into such numbers than is warranted.

Remember, AltaVista is designed primarily to help people find information. The counting is secondary, existing only because it is needed for the ranking function. For that purpose, it's okay if the estimate is off by a few thousand or more. Regardless of fluctuations in the results count returned, AltaVista will still return all of the best 200 matches for any given search—although how many hundred thousand "extra" matches it finds will vary.

During early tests in product development, AltaVista provided exact counts for all searches. Developers noted times in which 90 percent of AltaVista's resources were tied up counting while only 10 percent were answering queries. As a result, AltaVista now approximates most searches. Basically, if the number of matches is less than 200, the count is exact. And if the number is enormous—tens of thousands to millions—it's only a very rough estimate.

As soon as AltaVista determines that the number of documents to return for a given search will be large, it extrapolates the estimate from a partial result. And when AltaVista is busy, even that count may be truncated to let the system handle new queries. Because of this approach, the counts you get may vary when you repeat the same query. For instance, searching for a specific word at different times might yield a count as high as 190,000 or as low as 100,000.

In Advanced Search, you can put a check in the checkbox next to *Give me only a precise count of matches* to request a count. In that case, AltaVista makes its best effort to provide an exact count. But the system is always monitoring its own load, and if it decides that counting is slowing down others' searches, AltaVista will stop the exact count and provide a close estimate. For instance, someone might be tempted to gauge the popularity of Windows 95 over time by doing a search for *Win95* at regular intervals and comparing the counts. But since variations in the count might either be

random or reflect real information, there's truly no accurate method to assess the various counts that AltaVista returns.

In everyday terms, AltaVista's counting function is like the old joke about finding something—"Where did you find it?" In the last place I looked. "Funny, that's where I find stuff too." Just as few people continue looking after they've found the lost object, AltaVista doesn't work very hard at looking after it has found the results it will return to the searchers.

ABOUT THE ADVANCED SEARCH FORM

The Advanced Search Form is what you use to enter text and submit queries. You'll probably have to practice to figure out exactly what to enter in the Query and Ranking fields, but the examples later in this chapter should help you master this in no time.

The easiest way to build an Advanced Search query is to add information a little at a time—that is, add a term, try it out, then add a new term or phrase only if the search results aren't exactly what you need. You might end up with something that looks like a long, complex query, but because you built it piece by piece, you'll understand its exact function and it won't seem nearly so complex.

As you work through this chapter, you will become familiar with using the operators *and*, *or*, *not*, and *near*, as well as grouping parts of your search together with parentheses. Keep in mind that all the rules given in Chapter 2 for specifying capitalization, wildcards, and words and phrases remain the same here.

Tip *If at any time you need help doing an Advanced Search, click Help from the Advanced Search window. You'll be whisked directly to Advanced Search help, which shows you how to do Advanced Searches, discusses operators, and provides links to examples to learn from.*

About Operators

The words *and*, *or*, *near*, and *not* are called *operators*, which are what you use in Advanced Search to tell AltaVista how to interpret the query. For example, you can use *and* to create a query like *Pennsylvania AND 6500.* The operator *and* tells AltaVista to return all documents with *Pennsylvania* as well as with *6500.*

Note *When you combine search terms with operators, you create an expression. An expression can include any combination and any number of terms and operators.*

You can enter operators two different ways:

- You can type out the words—*and*, *or*, *near*, and *not*—using uppercase, lowercase, or some combination—whatever is easiest for you. We recommend, however, that you enter them in uppercase so that you can easily distinguish them from the search terms.
- You can use operator shortcuts, as shown in the following table.

Operator	Shortcut
AND	&
OR	|
NOT	!
NEAR	~

With the shortcut equivalents, you could either search for *weather AND forecast* or *weather & forecast.* Similarly, you could search for *skiing OR boating* or *skiing | boating*. Try out both ways, then use the method you find easiest—AltaVista doesn't care either way.

If you need to use a word that is also an operator as one of the search terms in an Advanced Search query, simply place that word in quotes. For example, suppose you're searching for information about Oregon and want to search for the word *Oregon* or its postal service abbreviation, *OR.* In this case, because the abbreviation is also an operator, you'd place the abbreviation in quotes, like this: *Oregon OR "OR."*

In the following sections, we'll tell you how each operator works and give you examples to help you learn to use them in your own Advanced Searches. As a starting point, look at the following table, which summarizes the operators and their functions.

Note *The following definitions aren't intended as a full course in Boolean logic—they're simply a starting point for use in Advanced Search.*

Operator	Function
AND	Ensures that both terms appear.
OR	Ensures that at least one of multiple terms appears.
NOT	Ensures that a term does not appear.
NEAR	Finds documents that include the search terms appearing within ten words of each other.

Including AND in Queries

Use the operator *and* to combine query terms or statements. You could enter a query like *bookshelves AND antique* to select documents with both terms. For that matter, you could keep adding *and* into your query to produce something like *bookshelves AND antique AND "for sale"* or even *bookshelves AND antique AND "for sale" AND "free delivery."*

Including OR in Queries

Use the operator *or* in queries to get documents that match either the first or the second term you enter. The results might match both words in your query, but they don't have to. You could, for example, search for *cat OR feline*. One, and possibly but not necessarily both, of the terms would be present in the resulting documents. Again you could string several *or* operators together to get a query like *cat OR feline OR kitten*.

Note *Don't forget to use operators. If you simply enter* cat feline, *AltaVista will try to find the phrase "cat feline" on Web pages, rather than searching for pages with either or both of the terms.*

Including NOT in Queries

Use the operator *not* to require that the term following it not be included in the results. For example, a search for *NOT interesting* will find all pages on the Web that do not include the term "interesting." Generally you'll use *and* with *not* to find anything matching the first term that also does not contain the second term. When you search for recipes on the Web, you could enter something like *recipe AND NOT liver* to obtain recipes without liver in them. AltaVista will not accept *not* alone to join two terms, so you'll get a syntax error if you enter a query like *meal NOT liver* (although we'd tend to agree with the sentiment).

Tip *You could search for something like* recipe OR NOT asparagus, *which would return all the documents containing the word* recipe *and all the documents in the entire Web that do not contain* asparagus. *That would be an enormous number of documents, but such a search is possible and could conceivably prove useful in special cases. If you find a good use for this search, let us know and we'll include your tip in the next version of the book. (We're still waiting for a good use for this one.)*

Including NEAR in Queries

Use the operator *near* to get documents that include both of the terms you entered within ten words of each other—preceding or following. For example, you might enter a query like *baby NEAR crib* or even something like *baby NEAR crib NEAR safe*. These queries would find documents that include the phrase "A baby laying in a crib is relatively safe," but not documents with "A safe baby will rarely be injured by normal household objects, although in special cases you may find that a defective crib can be hazardous." The difference between these two results is how close together the words appear. In the first result, the search terms appear within ten words of each other; in the second, they're more than ten words apart.

Tip *The operator* near *has no direct equivalent in Simple Search, although the proximity of terms is a factor when AltaVista ranks the results of Simple Searches.*

The *near* operator proves particularly helpful in searching for the names of people, which might appear in a wide variety of forms. Ideally, you'd search for *"Theodore Roosevelt"* and find all instances in which the name appears in that form. However, you might miss good pages because your search doesn't account for variations in how the name appears. While you could search for *Theodore AND Roosevelt* to find all documents in which both those words appear, your search would be far broader than necessary—matching instances when *Theodore* was at the beginning and *Roosevelt* was at the end, and there was no relationship whatsoever between those names (the former belonging to some other Theodore and the latter belonging to someone else with Roosevelt in their name). You could also try to imagine all the possible variants and enter them all in your query as alternative phrases connected by *or* (e.g., *"Theodore (Teddy) Roosevelt" OR "Roosevelt, Theodore" OR "Teddy Roosevelt"*), but that would be very tedious; and if you missed a variant, you might miss the very document you need.

If you instead use *near,* you'll find just about all the matches you need. The *near* operator also gives you a shortcut for narrowing your search when you suspect that two terms will probably be near one another. For instance, if you are trying to find the home page of the Acme Company—not just one of the dozens of Web pages that Acme might have—you can search for *Acme NEAR welcome* based on the assumption that most corporate home pages include the word *welcome* and place it close to the company name. (Of course, a search for *Acme NEAR coyote NEAR Wile* would be more fun because Wile E. Coyote—the roadrunner's nemesis—is mentioned on several Web pages.)

Tip *If you want to get really creative, you could search for* Acme AND coyote NEAR title:cartoon. *Yes, you can search by fields (such as title:) in Advanced Search, just like you can in Simple Search. See Chapter 2 for detailed information about searching by fields.*

Grouping Multiple Search Terms and Operators

As mentioned earlier in this chapter, Advanced Search (and Simple Search, to some extent) lets you combine any number of terms and operators to create expressions. You can easily string together a huge search expression with lots of terms and lots of operators, but for the results you need, you should probably use parentheses to make sure that AltaVista evaluates your expression correctly. Just as you (may have) learned in your math classes, expressions are evaluated in a particular order, not just left to right. 2+3*6 does not equal 30, because the multiplication has to happen before the addition; therefore, 2+3*6=20.

Likewise, AltaVista evaluates expressions in a specific order, not just left to right. Specifically, AltaVista evaluates operators in this order: *near, not, and, or*. The query *marketing OR sales AND business* would root out all pages with both *sales* and *business,* and then return all of those pages plus all the pages that match the word *marketing.* Generally, however, if you entered a search like this one, you'd really want all the pages with either *marketing* or *sales* and all of those pages that also contain the word *business.* Therefore, you'd want to use parentheses to produce *(marketing OR sales) AND business.*

When you include terms and operators in parentheses, you change the order in which AltaVista evaluates the query. Rather than evaluating the query according to the operator order, AltaVista evaluates expressions inside parentheses first, then moves outside the parentheses. So, in the Advanced Search query *(marketing OR sales) AND business*, AltaVista would find pages that include the terms *marketing* or *sales,* then work through those documents to find ones that also include *business.*

Or, if you have multiple parentheses in a query, AltaVista starts with the innermost parentheses, then works outward. Let's take a look at how you'd build a multi-parentheses query and see how AltaVista evaluates the terms and operators. Taking birds as an example, you could search for *robin OR oriole OR cardinal* to find documents with any of the terms. Unfortunately, you will also get some hits on things like the *St. Louis Cardinals* and the *Baltimore Orioles.* You could exclude anything with *St. Louis* and *Baltimore* or *baseball,* but a better way might be to group your terms together more effectively.

Because you want only information about these birds, you could search for *(robin OR oriole OR cardinal) AND bird* to find all documents with any of the three kinds of birds in addition to the word *bird.* Of course, you might need to include the technical term for the study of birds to make sure you get everything. Therefore, you might also search for *(robin OR oriole OR cardinal) AND (bird OR ornithology).*

If you find at this point that you're still getting unwanted sports references, you could add another set of parentheses and the expressions to exclude, like *((robin OR oriole OR cardinal) AND (bird OR ornithology)) AND NOT (baseball OR football).* This query will first find any documents with a mention of one of those three kinds of birds that also include the term *bird* or *ornithology.* It will then remove from the results list any of those documents that contain either the word *football* or *baseball.*

5

Note *Build a complex search like this gradually, rather than trying to type it in all at once. If you misplace or omit a parenthesis, or mismatch your parentheses, you might not get the results you want.*

As you can see, you have a great deal of control over your results when you use the Advanced Search. If you take a little time when constructing a query, you can help ensure that you get the information you need.

Tip *Remember that you can bookmark a search, and after you've created one of these monster search expressions, you should. Just create the query you want, submit it, and when the results page comes up, add a bookmark or add it to the Favorites list in your browser. When you return to the bookmark, your browser will automatically resubmit your query and provide the latest set of results to the parameters you specified.*

Taking Apart an Advanced Search

In the previous few sections, we've shown you how to build an Advanced Search using operators, terms, and parentheses. Let's take a quick look at what each component of an Advanced Search does. Suppose you want to find documents on American Indians. You would submit a query containing *American Indians* as well as *Native Americans* or even *tribe.* In Simple Search, you could simply list *"American Indians" "Native Americans" tribe*. In Advanced Search, you would have to join them together with *or* to get the same effect (i.e., *"American Indians" OR "Native Americans" OR tribe*). However, you can continue refining the Advanced Search using the operator *near* and parentheses to more accurately specify what you're looking for. The following table shows the evolution of this search.

Query	What You're Doing
American Indian	Basic concept for which to search.
American NEAR Indian*	The wildcard (*) ensures that plurals also show up, and using *near* broadens the range of possibilities by catching cases in which a few words separate American and Indian.
(American NEAR Indian*) OR (Native NEAR American*)	Add a second set of likely words (Native American) with *near* and a wildcard. Use parentheses to group the related terms. Using *or* between the parentheses ensures that either of the two elements shows up.
(American NEAR Indian*) OR (Native NEAR American*) OR tribe*	Add another single term.

Query	What You're Doing
(American NEAR Indian*) OR (Native NEAR American*) OR tribe* OR India*	Add an additional term. Now there are four possibilities, but the last one might also match terms such as India, the country. The next step will fix that.
(American NEAR Indian*) OR (Native NEAR American*) OR tribe* OR (India* AND NOT Asia*)	The final set of parentheses includes documents that match India* but do not also have the word Asia.

Note *This search used initial capitalization for many of the words. Only the capitalized version of these terms would be returned—not* American indian *or* native American, *for example. See Chapter 2 for more about how AltaVista treats capitalization in search queries.*

RESTRICTING YOUR SEARCH BY DATE

Advanced Search lets you limit your query to a specific range of dates by entering dates into the From and To fields, immediately to the right of the Ranking field. The dates you search by refer to the most recent revisions of Web pages—that is, the dates provided by the Web servers indicating when the documents were most recently changed. Unfortunately, there's nothing to indicate whether the whole document was overhauled on that date or a single typo was fixed. Despite that, you'll find that restricting searches by date is handy in cases like these:

- You have a rough idea of when something was placed on the Web.
- You are looking for facts and comments about an event.
- You don't want to bother with Web pages you consider too old—or too new—to be useful.

Note *Although restricting searches by date is a useful search technique, keep in mind that the date affixed to a Web page by the server is not always correct. For example, if the server's clock has the wrong date, the Web page will also have the wrong date. Although most dates on Web pages are accurate, you can't completely rely on them.*

Entering Dates

Entering dates is easy. Whether you want to enter a start date, end date, or both, all you have to do is enter *day/month/year*—in that order. Here are the guidelines to follow for entering dates:

- Use three-letter abbreviations (from most common European languages) for months.
- Use two-character abbreviations for years.
- Use as many characters as needed for days—one or two digits.

The following table shows sample dates in the correct format.

3/Jan/98	14/Feb/98	17/Mar/98
22/Apr/98	2/May/98	7/Jun/98
4/Jul/98	12/Aug/98	22/Sep/98
16/Oct/98	17/Nov/98	18/Dec/98

For your convenience, AltaVista provides a sample date in the From field, so you don't have to remember the format—just follow the model.

Tip *If you are using the date fields but aren't getting the results you expect, check to make sure that your start date is before your end date and that your start date is before today's date.*

Note *If you enter a date but omit the year, AltaVista fills in the current year. Likewise, if you enter a date but omit the month and year, AltaVista will complete both the current month and year. You do have to insert at least the day, however, if you want to search by date.*

RANKING WITHIN ADVANCED SEARCH

By default, Advanced Search returns results according to the search terms, operators, and parentheses you include in the query. Using the Ranking field, however, you can also specify how AltaVista ranks the matches it finds. So, for example, searching for *gardening AND vegetables* in Advanced Search with nothing in the Ranking field will give you lots of matching documents, in no particular order. You might get articles that only incidentally mention vegetables right up at the top, along with more topical articles about developing a vegetable garden, as shown in Figure 5-2.

Note *Using Advanced Search without including terms in the Ranking field rarely returns good results. The* only *exception to this is if you simply want to count or exhaustively list matches.*

If you want AltaVista to put the documents in a particular order, enter the more important terms in the Ranking field. You could simply put the terms from the Search field in the Ranking field (without the operators). That tells AltaVista to apply its usual ranking procedures based on those terms. The combination of terms in the Search field and the same terms in the Ranking field yields results comparable to Simple Search. In this case, adding *gardening vegetables* to the Ranking field gives more useful results than including no ranking terms at all would produce.

If you put different terms into the Ranking field from those in the Search field, however, you can rank and narrow your search at the same time. For example, if you put *beginner "getting started"* in the Ranking field (and keep *gardening AND vegetables* in the Search field), you'll get a much shorter and more precise list. AltaVista performs a second level of filtering and puts documents that contain the ranking words at the top of the list. With the Ranking field, just as with the search terms, the more information you provide, the more precise your results will be, as shown in Figure 5-3.

As a rule of thumb, always try several variants on your search and test your results to make sure that what you see is what you really wanted. It sometimes takes several tries to get the precise results you want, but once you have honed your search strategies, you can narrow your search more quickly in the future.

Translations | Browse by Subject | People Search | Business Search

Search the Web for documents in any language

gardening AND vegetables

Ranking: From: 21/Mar/86 To:

Give me only a precise count of matches.

search refine

Help . Preferences . New Search . Simple Search

About 15365 documents match your query.

Click to find related books at amazon.com

1. Summer Garden
Would you like to create a functional garden for your home or facility? Would you like to do it in a pinch? Do you need an outdoor recreational area...
http://www.missouri.edu/~projlife/garden.html - size 4K - 23-Apr-97 - English - Translate

2. The Greening of Rensselaer: Environmental Education Center - Environmental Ref
The EEC possesses many books and magazines for the Rensselaer community to use. We also have over 50 books on various topics. We will also update some of..
http://www.rpi.edu/locker/74/000874/EEC/eec_books.html - size 6K - 4-Mar-97 - English - Translate

3. Philadelphia Inquirer: Suburban North
Wednesday, August 6, 1997. Suburban North. Jewish educator drowns during visit to Israel Barbara Eidelman Wachs of Wynnewood was trying to rescue a boy...
http://www.phillynews.com/inquirer/97/Aug/06/pa_north/ - size 28K - 6-Aug-97 - English - Translate

4. Community Supported Agriculture Project
Do you want vegetables that dance? Spoutwood Farm Community Supported Agriculture Project. RD 3, Box 66 · Glen Rock, PA 17327 · (717) 235-6610. Pick up...
http://www.charm.net/~mroswell/csa.html - size 6K - 15-Aug-97 - English - Translate

5. SI: GARDENING AND ESPECIALLY TOMATO GROWING
GARDENING AND ESPECIALLY TOMATO GROWING. | Started By: Tomato Date: Apr 21 1997 12:58AM EST. A lot of people, me included, are in to gardening. I'm really.
http://www.techstocks.com/~wsapi/investor/subject-14596-Range-0 - size 18K - 15-Aug-97 - English - Translate

6. Club turns vacant lots into gardens bursting with blooms
Club turns vacant lots into gardens bursting with blooms. August 8, 1997. STEPHANIE SINCLAIR/Detroit Free Press%endphotocredit --> BY MARTY HAIR Free...
http://www.freep.com/fun/features/qlarge8.htm - size 9K - 8-Aug-97 - English - Translate

7. GARDENING HOTLINE
GARDENING HOTLINE. Get all the Answers Call: 1-900-562-1900 ext.3809 $2.49/min., Average call 3 min., 18 or older, Touch Tone Phone Required TeleService...
http://www.deltanet.com/allstar/garden.htm - size 5K - 15-Jun-95 - English - Translate

8. No Title
All contributions remain the intellectual property of the contributor/author. To reproduce any part of this site, in any form, be it via printed,...
http://web.ukonline.co.uk/Members/pat.fox/writer/story/st01.txt - size 21K - 15-May-97 - English - Translate

9. More About Stuart the Maniac
Hi! I'm Stuart the Maniac! Musical Cornucopia, CyberEvangelist, Linguophile, Internet and Spiritual Advisor. E-Mail address: maniac@io.com. Snail Mail...
http://www.io.com/~maniac/more.html - size 7K - 1-Jan-97 - English - Translate

10. No Title
GENERAL BOTANY. Return to Index Ordering & Payment Subject Selection (Home Page) A CHECKLIST OF THE ORCHIDS OF BORNEO Wood, J.J. Cribb, P.J. xii + 409 pp..
http://www.demon.co.uk/ssb/botgen.html - size 56K - 10-Nov-95 - English - Translate

1 2 3 4 5 6 7 8 9 10 11 12 13 14 15 16 17 18 19 20 ?

Our Network | Add/Remove URL | Feedback | Help
Advertising Info | About AltaVista | Jobs | Text-Only

digital Digital Equipment Corporation AltaVista

An unranked search

FIGURE 5-2

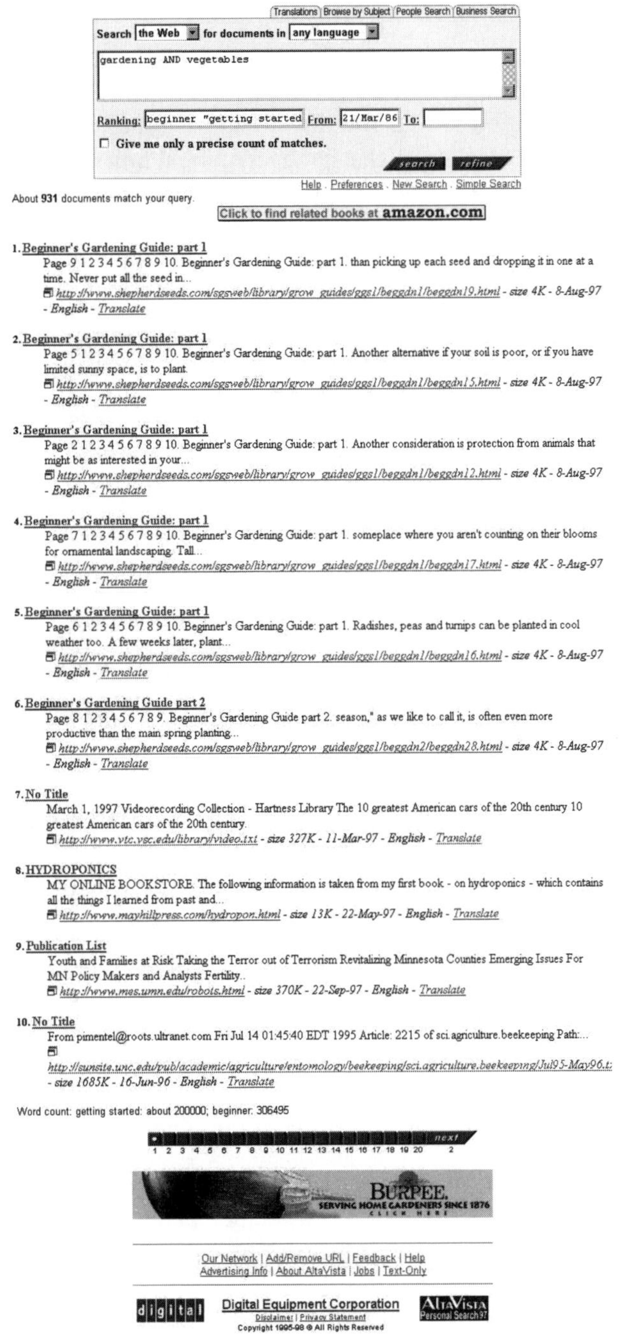

Translations | Browse by Subject | People Search | Business Search

Search the Web for documents in any language

gardening AND vegetables

Ranking: beginner "getting started From: 21/Mar/86 To:

Give me only a precise count of matches.

search refine

Help . Preferences . New Search . Simple Search

About **931** documents match your query.

Click to find related books at amazon.com

1. Beginner's Gardening Guide: part 1
   Page 9 1 2 3 4 5 6 7 8 9 10. Beginner's Gardening Guide: part 1. than picking up each seed and dropping it in one at a time. Never put all the seed in...
   http://www.shepherdseeds.com/sgsweb/library/grow_guides/ggs1/beggdn1/beggdn19.html - size 4K - 8-Aug-97 - English - Translate

2. Beginner's Gardening Guide: part 1
   Page 5 1 2 3 4 5 6 7 8 9 10. Beginner's Gardening Guide: part 1. Another alternative if your soil is poor, or if you have limited sunny space, is to plant.
   http://www.shepherdseeds.com/sgsweb/library/grow_guides/ggs1/beggdn1/beggdn15.html - size 4K - 8-Aug-97 - English - Translate

3. Beginner's Gardening Guide: part 1
   Page 2 1 2 3 4 5 6 7 8 9 10. Beginner's Gardening Guide: part 1. Another consideration is protection from animals that might be as interested in your...
   http://www.shepherdseeds.com/sgsweb/library/grow_guides/ggs1/beggdn1/beggdn12.html - size 4K - 8-Aug-97 - English - Translate

4. Beginner's Gardening Guide: part 1
   Page 7 1 2 3 4 5 6 7 8 9 10. Beginner's Gardening Guide: part 1. someplace where you aren't counting on their blooms for ornamental landscaping. Tall...
   http://www.shepherdseeds.com/sgsweb/library/grow_guides/ggs1/beggdn1/beggdn17.html - size 4K - 8-Aug-97 - English - Translate

5. Beginner's Gardening Guide: part 1
   Page 6 1 2 3 4 5 6 7 8 9 10. Beginner's Gardening Guide: part 1. Radishes, peas and turnips can be planted in cool weather too. A few weeks later, plant...
   http://www.shepherdseeds.com/sgsweb/library/grow_guides/ggs1/beggdn1/beggdn16.html - size 4K - 8-Aug-97 - English - Translate

6. Beginner's Gardening Guide part 2
   Page 8 1 2 3 4 5 6 7 8 9. Beginner's Gardening Guide part 2. season," as we like to call it, is often even more productive than the main spring planting...
   http://www.shepherdseeds.com/sgsweb/library/grow_guides/ggs1/beggdn2/beggdn28.html - size 4K - 8-Aug-97 - English - Translate

7. No Title
   March 1, 1997 Videorecording Collection - Hartness Library The 10 greatest American cars of the 20th century 10 greatest American cars of the 20th century.
   http://www.vtc.vsc.edu/library/video.txt - size 327K - 11-Mar-97 - English - Translate

8. HYDROPONICS
   MY ONLINE BOOKSTORE. The following information is taken from my first book - on hydroponics - which contains all the things I learned from past and...
   http://www.mayhillpress.com/hydropon.html - size 13K - 22-May-97 - English - Translate

9. Publication List
   Youth and Families at Risk Taking the Terror out of Terrorism Revitalizing Minnesota Counties Emerging Issues For MN Policy Makers and Analysts Fertility..
   http://www.mes.umn.edu/robots.html - size 370K - 22-Sep-97 - English - Translate

10. No Title
   From pimentel@roots.ultranet.com Fri Jul 14 01:45:40 EDT 1995 Article: 2215 of sci.agriculture.beekeeping Path:...
   http://sunsite.unc.edu/pub/academic/agriculture/entomology/beekeeping/sci.agriculture.beekeeping/Jul95-May96.t: - size 1685K - 16-Jun-96 - English - Translate

Word count: getting started: about 200000; beginner: 306495

1 2 3 4 5 6 7 8 9 10 11 12 13 14 15 16 17 18 19 20 next 2

BURPEE. SERVING HOME GARDENERS SINCE 1876 CLICK HERE

Our Network | Add/Remove URL | Feedback | Help
Advertising Info | About AltaVista | Jobs | Text-Only

digital Digital Equipment Corporation
Disclaimer | Privacy Statement
Copyright 1995-98 © All Rights Reserved
AltaVista Personal Search97

A ranked and precisely focused search

FIGURE 5-3

Why You Might Not Want to Rank Results

Although ranking results is handy for most queries, there are instances when you might not want to use the ranking feature. If your purpose is to gather a set of relevant but unordered information, you might choose not to rank the results. For example, if you want a list of all pages that have a hypertext link to your Web site—information in which every match is equally important—you probably would not want to rank the results. You can also use unranked searches to survey the body of knowledge on the Web about a specific topic. The main reason for choosing not to rank is to be able to view a larger number of matches.

If you use Advanced Search and choose not to rank, you can keep hitting the Next button at the bottom of the results page over and over again, seeing many more matches—for example, if you want a long list of Web pages that have hypertext links to your company's pages.

Note *The reason ranked searches only return the top 200 matches is that it takes AltaVista quite a bit of time to calculate rankings. By ranking and returning only the top 200 results, AltaVista can move on to other work and get your results to you more quickly. Also, the benefit to be derived from ranking more than 200 results would be minimal, because 76 percent of users only look at the first page of query results (the top ten matches), and less than 5 percent ask for more than five screens (fifty items) of results.*

TROUBLESHOOTING ADVANCED SEARCHES

Completing a successful Advanced Search takes some practice. You might not get the results you expect right away, but using the following troubleshooting steps should quickly improve your results.

- *If your results don't look as promising as you'd hoped,* first take a look at the word and document counts that are provided above the list of documents to check whether you entered the search syntax (*and, and not,* wildcards, parentheses, etc.) as you'd intended.
- *If you expected all of your search terms to show up in every document and they didn't,* verify that you used the operator *and.*
- *If you expected words of a phrase to appear together and they didn't,* make sure you put quotation marks around the phrase.

- *If your results seem incomplete* and you used punctuation marks or capital letters, try omitting the punctuation and changing the capital letters to lowercase.
- *If you get Syntax error (Bad Query)* and you were using complex constructions with parentheses, try checking your parentheses and make sure they are all in pairs (there are the same number of left and right parentheses). If that doesn't help, try simplifying the query and adding terms and expressions back in gradually.
- *If you're getting too many documents to sort through,* review the results from your original search and see if the documents that don't interest you have anything in common—a word, a part of their URL, or any element that you could use to exclude them from future search results by using *and not* in Advanced Search. Additionally, check the ranking terms you entered to verify that you're ranking according to the right terms.
- *If a document you found just doesn't look like it fits with the rest,* keep in mind that the Web is constantly changing. Some people edit their pages frequently. It is possible that when the page was retrieved and indexed it would have been a match for your query, but the new version (which hasn't yet been re-indexed) doesn't fit. Eventually the old information will be dropped from the index, and the new content will be indexed. But for now, you are left with a mismatch.

BRINGING IT ALL TOGETHER: A SAMPLE ADVANCED SEARCH

Now that you've seen all the different things you can do with Advanced Search—from using operators (*and, or, near, not*) to restricting searches by date to using ranking criteria—here's an example of how to pull it all together.

For the sample Advanced Search, let's take a look at a search we did for sample business plans, which you might find handy for developing your own business plan. This example started with a regular Advanced Search query: *"business plan" OR "marketing plan."* Then, we added the word *sample* to the Ranking field to try to ensure that a usable example would be found.

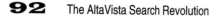

Translations | Browse by Subject | People Search | Business Search

Search the Web for documents in any language

"business plan" OR "marketing plan"

Ranking: sample From: 21/Mar/86 To:

Give me only a precise count of matches.

search refine

Help . Preferences . New Search . Simple Search

About **10433** documents match your query.

Click to find related books at **amazon.com**

1. **Sample Book Outline (Extended)**
   Sample Book Outline (Extended) This is a sample detailed (extended) book outline, in the exact format I've used to sell more than 30 non-fiction books. I..
   http://w3.one.net/~banks/sampleb2.htm - size 14K - 5-Jun-97 - English - Translate

2. **Value Added Information Business: cynoB (Sample)**
   An opportunity pack (OppPak) that combines Information Broking and Knowledge Engineering. Illustrated by a Strategic Plan presented as an HTML (web)
   http://ourworld.compuserve.com/homepages/DecisionAid/Opak2.htm - size 27K - 22-Aug-97 - English - Translate

3. **Sample Course Syllabus- Part 2**
   Sample Course Syllabus- Part 2. What to look for in a team. Here are a few things to look for as you review each team and assess the risk of investment:...
   http://www.acad.utk.edu/marketplace/mk02006.htm - size 10K - 7-May-97 - English - Translate

4. **BANK TECHNOLOGY NEWS - Sample Issue**
   BANK TECHNOLOGY NEWS via NewsNet. Sample issue headlines from JANUARY, 1996. Visit http://www.newsnet.com for more details.
   http://www.newsnet.com/libiss/fi70.html - size 137K - 21-Apr-97 - English - Translate

5. **MODERN PLASTICS - Sample Issue**
   MODERN PLASTICS via NewsNet. Sample issue headlines from January 1996 . Visit http://www.newsnet.com for more details.
   http://www.newsnet.com/libiss/ch23.html - size 226K - 21-Apr-97 - English - Translate

With these initial results, we didn't find many good samples of business plans in the first few matches. So, we decided to revise the search query to include either *business* or *marketing* and the word *plan*. The revised query became *(business OR marketing) AND plan.*

This search generated lots of matches, but they still seemed pretty far afield. The *and* operator in the query just required that both words appear in the document. In this example, we shifted the word *sample* from the Ranking field to the Query field and moved the phrase *"business plan"* to the Ranking field. The query was now *((business or marketing) NEAR plan) NEAR sample* and in the Ranking field was *"business plan."*

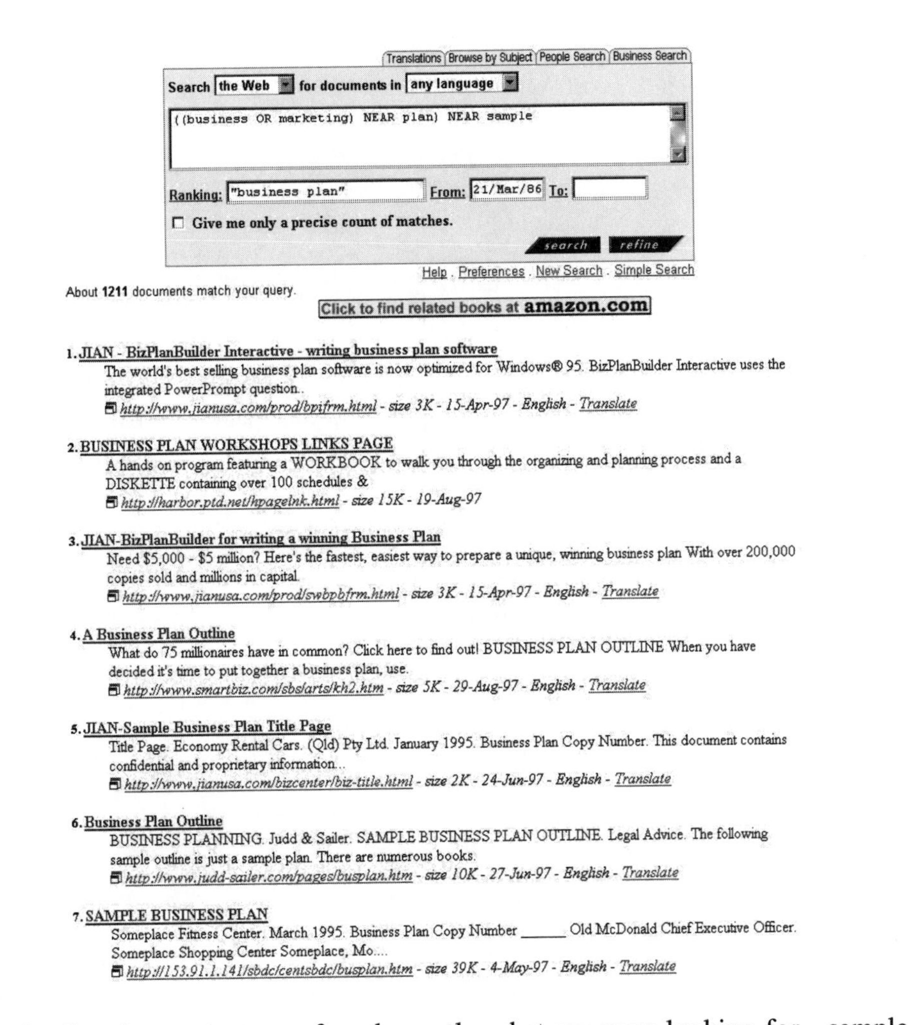

In the first few screens we found exactly what we were looking for—sample business plans.

Because business plans are relatively timeless, we didn't attempt to narrow the search with dates. Additionally, we didn't actually have to fine-tune the search as much as we did in this example. Refining the search as we did yielded good results within the first few results pages. Had we stopped sooner, we could have found many of the same matches, but would have had to look through more results pages.

Note *Don't forget! You can improve your results by searching according to the fields* text, title, link, anchor, url, host, domain, image, *and* applet. *Just enter the field name followed by a colon and the word or phrase you want. Refer back to Chapter 2 for more information about searching by field.*

What's the Difference? Comparing Advanced Search with Simple Search

Although Simple and Advanced Search have some similarities, they are in fact quite different in how they handle the four operators, punctuation, and rules for words, phrases, and capitalization. This section shows you how the two search methods differ.

Operators

Three of the four Advanced Search operators—*or, and,* and *not*—have close equivalents in Simple Search. This table presents these operators and the following section explains how the equivalents work.

Advanced Search	Simple Search
OR	Entering terms with only a space between them
AND	Placing a + before both the terms that in Advanced Search are linked by AND
NOT	Preceding a term with -

When you connect two or more words or phrases with *or* in Advanced Search, such as *Macintosh OR Mackintosh*, AltaVista processes the query as it would *Macintosh Mackintosh* in Simple Search (except that Simple Search ranks the results and Advanced Search does not). In other words, this Advanced Search with *or* is the same as using Simple Search and entering a series of terms with no commands at all. Such a query matches all documents that contain any one of the words or phrases, but not necessarily all of them.

You cannot explicitly use the operator *or* in Simple Search. You won't get an error if you try, but it will just be interpreted as a word and then be discarded because it is so common. You only need to type the important words in Simple Search. On the other hand, in Advanced Search, you cannot simply type a list of words and phrases; you must connect them with an operator such as *or*—otherwise they'll be interpreted as a phrase.

When you connect two or more words or phrases with *and* in Advanced Search, such as *governor AND Nevada*, your query is roughly equivalent to *+governor +Nevada* in Simple Search, except that the Simple Search results will be ranked. In other words, all those words or phrases must be present in a document for it to count as a match.

When you connect two or more words or phrases with the operators *and not* in Advanced Search, such as *digital AND NOT watch,* your query is equivalent to *digital -watch* in Simple Search. In other words, the word or phrase preceded by *and not* or the minus sign must be absent for a document to count as a match.

Remember, if in Advanced Search you want to use any of the four operator command words as words rather than commands, you have to put them inside quotation marks. For instance, you might want to search for *"OR"* as an abbreviation for Oregon. In that case, you might search for *Oregon OR "OR."*

Rules for Queries

In Advanced Search, the rules for defining words and phrases, capitalization, wildcards (*), and structure are the same as for Simple Search. But there are instances where the same expression in a query could have a different meaning in the two search modes. For instance, in Advanced Search, the symbols + and - are interpreted as punctuation; in Simple Search, + and - are operators that carry special meaning for AltaVista. However, the punctuation characters &, |, !, and ~ have meaning in Advanced queries but not in Simple Search.

Punctuation

In Advanced and Simple Search modes, most punctuation marks serve the double function of indicating the beginning or end of a word and linking a string of words together as a phrase. However, in Advanced Search mode, parentheses and the symbols used as commands (& | ! ~) are interpreted differently. To avoid confusion, if you want to create a phrase in Advanced Search, simply enclose the text in quotation marks.

SOME POINTERS TO OTHER USEFUL CHAPTERS

In this chapter, you learned the ins and outs of Advanced search—and we bet it was much easier than you thought! After you've mastered Advanced Search, you might check out one of these handy chapters:

- Visit Chapter 4 and learn how to use AltaVista's incredible Refine feature.
- Refer to Chapter 6 to find out how to search Usenet (also known as network news).
- Cruise through Chapter 9 for examples of how other people use AltaVista to create effective searches.

See you there!

Searching Usenet Newsgroups

Use AltaVista to search Usenet newsgroups! *Usenet newsgroups* (sometimes called just *Usenet*, *news*, *network news,* or *newsgroups*) are collections of messages addressed to a group of people with common interests, rather than simple email messages addressed to a single individual. Each of the over 20,000 newsgroups in existence has a designated subject, on any topic from books to the culture of Nepal to obscure variations of UNIX.

While Web pages tend to be carefully constructed, edited, and formatted, newsgroup postings are typically informal, spontaneous, interactive messages. Although some newsgroups are moderated—meaning that someone filters messages for relevancy and usefulness before they get posted to the newsgroup—most newsgroups are unmoderated, giving you uncensored opinions and insight from people who are familiar with (or at least interested in) the topic at hand.

In this chapter, you'll learn the benefits of searching newsgroups with AltaVista and how to do Simple and Advanced newsgroup searches. You'll also see how to focus your newsgroup searches to make them as productive as possible. Plus, we'll show you how to ensure that you find just messages that meet particular conditions you set and some special applications for newsgroup searches.

About Newsgroup Categories

You'll find loads of newsgroup categories that include a wide range of topics—all of which stemmed from seven original categories:

comp.	Addresses anything computer-related.
rec.	Includes topics like hobbies or pastimes.
news.	Deals with the Usenet news network itself.
sci.	Deals with science. Surprise!
soc.	Addresses a variety of social issues, often related to a particular culture.
talk.	Encourages debate and expressing opinion for the fun of it.
misc.	Addresses almost anything else and includes such useful groups as misc.jobs and misc.forsale.

As newsgroup popularity grew, many other categories were formed. These other categories originated in a variety of ways but are now part of the same newsgroup environment. You can get all of these and more through AltaVista, but your local news server might or might not provide access to the whole collection. Some of the most common additional categories include the following:

alt.	Covers every subject under the sun. These groups all bypassed the traditional newsgroup creation process for a variety of reasons.
bionet.	Addresses biological sciences.
bit.	Reproduces messages from some of the popular email distribution lists, usually known as listservs.
biz.	Welcomes advertising and marketing, unlike most other groups.
de.	Covers a full range of topics, all in German.
fj.	Covers a full range of topics, all in Japanese.
ieee.	Covers topics for the Institute of Electronic and Electrical Engineers.
gnu.	Is for the Free Software Foundation and its GNU project.
k12.	Is intended for topics about education from elementary through high school.
vmsnet.	Focuses on Digital Equipment's VAX/VMS operating system.

Additionally, you'll find a whole slew of other newsgroups targeting geographic areas, which are evident in the newsgroup category names, such as the following:

ne.	New England
ba.	San Francisco Bay Area
utah.	Utah
ok.	Oklahoma

WHY SEARCH NEWSGROUPS WITH ALTAVISTA?

Though you could subscribe to and read any number of newsgroups and get the information you need, searching newsgroups through AltaVista offers you several benefits. First, with millions of total news items posted every day, very few people can find the time to keep track of what is happening in more than a few groups. AltaVista allows you to search through all the words in all the current postings to all of the newsgroups and immediately read just the items you choose. Rather than subscribing to multiple newsgroups and wading through hundreds (or thousands!) of messages, all you have to do is type in a few words and scan AltaVista's results list to find the information you need.

Second, searching newsgroups provides you with the most recent information available on a topic. Unlike Web pages, which stay in AltaVista's index indefinitely, newsgroup postings are deleted after some period of time—usually two to four weeks. Depending on disk space and message volume, some messages will be retained as long as several weeks, while others might expire from the server in only a couple of days. After a couple of months, however, the oldest news postings certainly have expired off of the AltaVista server, only to be replaced by a fresh batch. The result is that you always have access to the most recent information, without having to wade through outdated postings.

Third, AltaVista provides complete access to newsgroups. AltaVista not only carries a full "feed" of over 20,000 newsgroups, but also lets you directly access each individual message among the millions posted each day.

Normally, to read newsgroups you must have access to a news server, and you must know how to use your news reader software. Even then, you might find that, for a variety of reasons, your news server handles only a very limited selection of groups. By searching newsgroups through AltaVista, you can search the full collection of newsgroups and get the full text of all articles.

Along those same lines, you can search for the specific information you need—searching on the conditions you set. As an example, think of the benefits of being able to search for all messages that mention your company. Finding out what your customers, competitors, or business partners are saying about your company could be useful, couldn't it? Or consider using AltaVista to locate posts or messages by an individual or those that include a specific phrase. You can search by the author of a posting and find everything else that he or she has written to any newsgroup over the last month or so.

Tip *Some hiring managers search AltaVista for postings by potential employees—just to see what they're saying, how, and to whom.*

Fourth, searching newsgroups through AltaVista eliminates having to wade through spam. Many—probably most—newsgroups are cluttered with off-topic or irrelevant messages designed to get readers to buy or sample a product or service. These messages, called *spam*, make it difficult or frustrating to follow specific discussions because the electronic junk mail gets in the way. If you're searching for newsgroup postings on a specific topic, however, the spam is unlikely to show up as a match, and, therefore, you're less likely to see it at all.

Finally, AltaVista allows you to construct a very focused search and find items on that topic in dozens of different groups that you might otherwise not have known existed. This helps you discover previously unknown groups that you might want to read and participate in regularly. Likewise, when you have something to say or ask, you can quickly find newsgroups where it would be appropriate to post your remarks.

SEARCHING NEWSGROUPS

Searching newsgroups is very similar to doing Web searches—in fact, many of the same principles and concepts you learned in Chapters 2 and 5 apply here, too. Let's take a look at doing Simple and Advanced Usenet searches. Before you begin, go ahead and connect to the Internet, open your browser, and access AltaVista's home page by typing in **http://altavista.digital.com**.

Remember *When a query yields good results, you can bookmark that query; the next time you click that bookmark, you can resubmit the search and get fresh results.*

Doing a Simple Search of Newsgroups

When you first arrive at the AltaVista home page and Search Form, the drop-down menu at the left shows the default selection, "the Web." Use the drop-down menu to tell AltaVista to search Usenet, rather than the Web, as shown here:

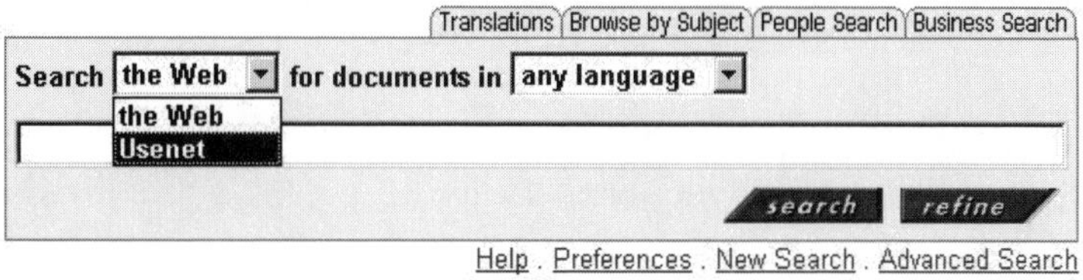

Translations | Browse by Subject | People Search | Business Search
Search the Web for documents in any language
the Web
Usenet
search refine
Help . Preferences . New Search . Advanced Search

So, a sample Usenet search might look like this:

1. Click the down arrow and choose Usenet.
2. Type **vacation** in the Search field.
3. Press ENTER or click the Search button.

Almost immediately, you'll see a list of thirty items that match the search terms you entered. The response is virtually instantaneous, even though you just searched through several million current newsgroup postings.

Note *The Language and Refine options are not currently available in Usenet searches, due to differences in the way the Usenet index and Web index are constructed.*

Doing an Advanced Search of Usenet

Just as the Simple Search of Usenet was like the Simple Search of the Web, Advanced Search of Usenet is like Advanced Search of the Web in many ways. You should submit Advanced Usenet searches if you want to do any of the following:

- Restrict your search based on dates
- Search for a word or phrase that occurs *near* another one
- Organize long, complex queries
- Retrieve more than 200 matching items

To do an Advanced Search of Usenet, breeze through the following table for a quick reminder of Advanced Search operators. Then use the following steps, which take you through a sample Advanced Search of Usenet.

A Quick Reminder about Operators

Operators, discussed in detail in Chapter 5, are what you use in Advanced Search to tell AltaVista how to interpret a query. The following table reviews the operators and their equivalents, and provides a brief explanation of each one.

Operator	Shortcut	Explanation
AND	&	Use AND to make sure that both terms are present in the resulting documents.
OR	|	Use OR to get documents that match either the first or the second term.
NOT	!	Use NOT to exclude something.
NEAR	~	Use NEAR to get documents that include both terms within ten words of each other.

1. Choose Advanced Search from the main AltaVista page.
2. In the Advanced Search form, click the down arrow beside Search and choose Usenet.

Translations | Browse by Subject | People Search | Business Search

Search Usenet for documents in any language

(vacation AND inexpensive) NEAR mountain*

Ranking: From: 21/Mar/86 To:

Give me only a precise count of matches.

search refine

Help . Preferences . New Search . Simple Search

3. Type **(vacation AND inexpensive) NEAR mountain*** in the Search field.
4. Click the Search button.

Just as with Simple Search, you'll almost immediately see a list of up to thirty items, culled from the millions of newsgroup postings.

ABOUT NEWSGROUP SEARCH RESULTS

The results from newsgroup searches, shown in Figure 6-1, are similar to the results from a search of the Web with Compact results. (For more information about seeing Web results in Compact Form, see Chapter 2, or read up on AltaVista preferences in Chapter 7.)

Compact results are simply results compressed into one line. Following is an explanation of what each segment of the Compact results means.

Documents

The Documents line tells you the number of documents that AltaVista found on your search. When you first submit your search, you will have documents numbered 1 to 30, generally out of a much larger number of documents. AltaVista may also have found duplicate postings—items that were "cross-posted" to several different groups at the same time—and will report back if such duplicates were discarded. Because of the discarded duplicates, some screens won't actually show 30 entries—you'll see 30 entries minus the duplicates.

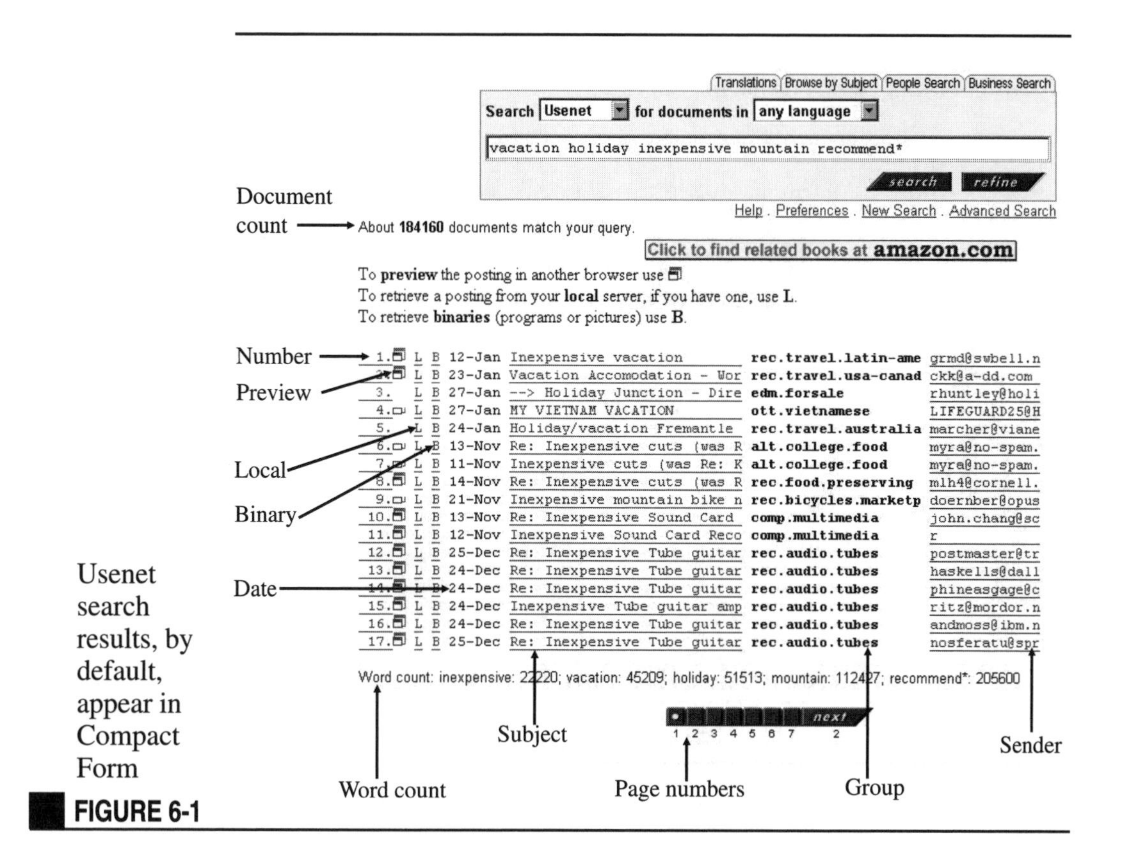

Usenet search results, by default, appear in Compact Form

FIGURE 6-1

Number

The first column numbers each match in order, just to make it a little easier for you to keep track of which items you're going to look at—seventeen different messages with "Re: No, you're wrong" kinda blend together otherwise.

Preview

The Preview button lets you open up the match in another browser window. If you select this option when your browser window is maximized, you may need to look through the open browser windows on your system to find the Usenet posting. You'll

find this handy, though, because you can keep the results visible at the same time you're looking at one of the messages.

Local (L)

The letter *L* stands for *local.* If you click it, you connect to your local news server (if you have one and have selected it in the setup of your browser) to read the item there. If, for example, you live in Montreal and have a slow Internet connection, it would likely be better to connect to your local news server, rather than reading news from the AltaVista server in Palo Alto, California. Local connections, by their very nature, are faster than long-distance ones.

If you're using the local option, you may not be able to retrieve the posting from your local server, for one of the following reasons:

- Your local server may not include the newsgroup you're interested in.
- Your local server's expiration policy may not be the same as AltaVista's.
- Your local server may not yet have received the article.

In all cases, if the local option doesn't work, you can still retrieve the posting directly from AltaVista.

Binary (B)

The letter *B* stands for *binary mode*. Normally, AltaVista will reformat the text in a newsgroup posting into HTML code for viewing on your browser—including putting matching terms in boldface. This results in clear, bold headings and pages formatted so they look more readable in your browser. Some newsgroup postings, however, contain encoded binary information, such as clip art files or executable programs, and the Binary option lets you download them in a usable and decodable form, which the regular link will not provide.

Date

The Date column tells you the date as a number and abbreviation for the month. You'll notice that the year isn't visible. It really isn't necessary, because newsgroup postings expire as they age and are deleted when they expire—in six to eight weeks, maximum.

Subject

The Subject column shows the subject of the posting. The subjects generally, but not always, reflect the content. Remember, sometimes people change the topic of the message without changing the subject line to correspond to what they're talking about. Just click the subject to have AltaVista deliver the full text of the item to you immediately.

Newsgroup

The Newsgroup column shows the newsgroup name—or part of it. Longer newsgroup names are truncated, but you'll still see enough of the name to tell if the group may be of interest to you.

Sender

The Sender column shows the email address of the person who posted the message. If you click on the address (and your browser is configured correctly, and the sender set up their email address correctly), you'll be able to send email directly to that person from your browser.

As problems with unsolicited commercial email messages (often called spam) become more prevalent on the Internet, you'll find that many of these addresses don't really work—they'll say something like *someid@removethiswordand theperiodtoreply.raycomm.com*. In these cases, you'll have to read the address or instructions in the body of the message to send a message back to the original poster.

Page Numbers

Page numbers, along with Prev(ious) and Next links, are visible in the footer of the Search Results. The yellow dot shows you which page of results you're on. Click the individual buttons to see additional pages of results, or click Next to move to the next page. After you've moved to one of the other pages of results, you'll have the additional choice of Prev, which takes you to the previous results pages. Note that you can always link directly to all other pages of results.

Word Count

The Word Count line tells you how many times your search term or terms were found. Additionally, if you searched for a particularly common word, the Word

Count line will tell you which terms were so common that they were discarded for purposes of ranking the relevance of particular documents.

Footer

The footer information is just like that of the regular Simple Search results—it presents links to a variety of information about AltaVista and Digital Equipment Corporation.

Remember *You can use the AltaVista Preferences page to select Detailed format for your results. You'll see the same basic information, but in a more complete and less cryptic format, just like the regular Web search results. See Chapter 7 for details.*

Getting a Detailed Results List

If you can't tell whether the results are what you had hoped for from the Compact format, visit the Preferences page and choose Detailed view for Usenet search. Detailed results include the same basic information, but in a format more closely resembling Web search results, giving you a less telegraphic view of the results, but still no summaries of the documents. See Chapter 7 for more about setting Preferences.

FINE-TUNING USENET SEARCHES

Fine-tuning Usenet searches is a bit different from fine-tuning Web searches. First, because you cannot use Refine to help make sure you see what you want, you'll need to be somewhat more precise with your actual search terms. Second, because of the different tone and style of Usenet in general, you may find that a great Web search yields less than satisfactory results when applied directly to Usenet. So, what can you do to improve Usenet search results? Experiment and practice a little, and try some of the tips in this section.

Fine-Tuning Simple Usenet Searches

You can refine Usenet searches in the following ways:

- Add words to the search query. Suppose that your new puppy is chewing everything in sight. Rather than entering the search query *puppy,* you might try *puppy training* or *puppy chewing*.
- Require words to appear in search results by using the plus (+) sign. For example, you could try *+puppy +chewing* to require that both terms appear in results documents.
- Eliminate words from search results by using the minus (-) sign. For instance, you could eliminate chew toys from search results with the query *+puppy +chewing -"chew toys."*
- Use wildcards. You could narrow your search down by adding *chew** and requiring all the words to be present. For example, *+puppy +training +chew** will get you a full collection of people's opinions on the subject, with any variant on chew, chewed, chewing (or for Star Wars fans, Chewbacca).
- Search by newsgroup components. For instance, you can enter *subject:chewing* in the query field to find postings with chewing in the Subject line.

Fine-Tuning Advanced Usenet Searches

In addition to using the by-now-familiar Advanced Usenet searches with operators, you can also fine-tune them by searching by date and by using AltaVista's Ranking option.

Fine-Tuning by Date

As with Web searches, to search by date, simply enter a date in the From field and the To field. For example, if you put today's or yesterday's date in the Start field, your query will be restricted to the most recent articles available on AltaVista. Searching by date usually proves particularly useful in Usenet, simply because of the volume of posts and the speed with which they change. Here are some guidelines to remember about searching by date:

- Enter the date using the *day/month/year* format, such as 31/Dec/93. The day and year are both numbers, but the month must be the three-character abbreviation, as shown here:

Jan Feb Mar Apr May Jun Jul Aug Sep Oct Nov Dec

- Make sure that your Start date is before your End date and that your Start date is before the current date.
- Remember that if you enter a date but omit the year, AltaVista fills in the current year. Likewise, if you enter a date but omit the month and year, AltaVista will add the current month and year. You do have to insert at least the day if you want to search by date.
- Remember that Usenet postings are periodically deleted. Don't expect to find items more than a few weeks old, although the actual length of time that AltaVista keeps specific newsgroup articles varies.

Fine-Tuning by Ranking Results

Another way to refine Advanced Usenet searches is to include information in the Ranking field. While no one method will work for all cases, we've found that Usenet searches tend to work well if you use a relatively general term for the search, then augment it with a very specific Ranking term.

For example, to find woodworking patterns, search on *woodworking,* then put *patterns* in the Ranking field. The results are more focused and useful than a search for *woodworking and pattern.*

Alternatively, search for *cookies,* then rank by *chocolate chip.* You'll get cookie recipes, with *chocolate chip* ones at the top, then *chocolate* or *chip* lower in the list. (No, we don't know what a "chip cookie" might be, and don't care to speculate.)

Fine-Tuning by Searching for Fields

Finally, in both Simple and Advanced Usenet searches, you can fine-tune searches by searching for *fields*, which are just parts of the posting, such as the To:, From:, and Subject: lines

Here, instead of searching for a particular topic by entering terms, you tell AltaVista to search for terms within the field you specify. For example, you could search for specific content in the Subject: line or search for addresses in the From: line. The following table summarizes fields you can search by.

Fields	At-a-Glance Description	How to Enter the Search Information
from:	Searches only in the From field. Matches email addresses and sometimes real names or company names.	from:bill@whitehouse.gov from:"Bill Clinton"
subject:	Searches news articles' subjects. You can combine this with a word or phrase.	subject:"for sale" subject:election
newsgroups:	Matches news articles posted in newsgroups with that name (or partial name). Often used in combination with other search terms.	newsgroups:rec.humor
summary:	Searches in the Summary field (of articles that have summaries).	summary:invest*
keywords:	Searches through the keywords of all articles that have keywords.	keywords:NASA

from:

The *from:* field gives you the email address, the full name of the poster, and sometimes other identifying information, such as a nickname or employer name. Generally, you'd search for the whole or partial email address, as in *from:president@whitehouse.gov*.

You could also search for *from:magna.com* to limit your search to postings by anyone with *magna.com,* for instance, as the host name of his or her mail address. A search like *from:ford.com* would limit your search to postings by employees of Ford Motor Company (posted from their corporate address), which could be quite useful if Ford is a competitor or customer of yours. Anything that might appear in the From: line of a posting is fair game for search. You could easily search for

from:"Lee Iacocca" and find all of his postings to newsgroups (if there are any), or combine a *from:* search with other search terms to get a more precise listing.

Tip *Keep in mind that email addresses can easily be faked—just because you found a post from bill@whitehouse.gov doesn't mean that Bill Clinton really sent it.*

subject:

The *subject:* field lets you search for anything that might appear in the subject line. If you are looking for information on a product called the "Gooblefitz" and would like to focus on those articles that have that as the main topic, use *subject:Gooblefitz* as a search parameter.

Searching by subject line works no matter how many other words appear in the subject line and whether the term you specify is first, last, or in the middle. The element *subject:* is also a good way to track down an entire discussion on a single subject, including replies and forwards, because the subject line of a reply is typically in the form of *re:"subject name of the original posting."*

newsgroups:

The *newsgroups:* field limits the search by newsgroup. You can submit a query that uses the entire name of a group, as in *newsgroups:rec.arts.books*. You can also open your search to a large category of newsgroups, such as, in this example, *newsgroups:rec*.

Tip *Don't be surprised if you search for only a specific newsgroup but you see AltaVista list messages from newsgroups you didn't specify. This happens because a message may have been posted to your group and to others, but AltaVista displays the name of only one of the "cross-posted" newsgroups.*

About Newsgroup Names

You might recall from the "About Newsgroup Categories" sidebar, at the beginning of this chapter, that newsgroup categories are named with cryptic abbreviations. These category names are built from a series of abbreviations, separated by periods, that in their left-to-right sequence indicate a hierarchy from general to specific. For example, within the rec. category, you find, among others, woodworking and food as subcategories. Within the food subcategory, you'll find items such as rec.food.cooking, rec.food.baking, or rec.food.recipes. The subsubcategories (cooking, baking, and recipes) belong within the food subcategory, which belongs within the general recreation (rec.) category.

You can use this structure, along with the *newsgroups:* field, to readily limit potential matches to useful areas. For example, if you're looking for technical information, you can probably safely exclude most alt. groups, or if you're looking for hobby information (and you're not into computers), you can exclude the comp. group and include the rec. and alt. groups.

summary: and keywords:

The *summary:* and *keywords:* fields provide specific additional information about the posting. Whether you can find information from these fields depends on whether the author provided a summary or keywords. The author of a newsgroup posting may, depending on the software used, have the option to attach a synopsis and to designate certain words as keywords, to help people with news reader software to find the posting and save time figuring out if it's something they should read. At the time of this writing, few authors provided these organizational tidbits, though.

SOME USEFUL USENET SEARCH APPLICATIONS

Usenet searches have additional uses beyond the fairly obvious ones of searching for subjects that interest you. AltaVista makes it easy to really put the power of Usenet to work, such as:

- Finding responses to your own postings
- Filtering postings
- Keeping your postings from being found

Finding Responses to Your Usenet Postings

If you are an active newsgroup participant, you might find that tracking responses to your postings is difficult. For example, if you post a number of different items to a number of different groups, you have to check each of those groups on a regular basis, which can be very time consuming. Using AltaVista, however, you can quickly and easily find any replies to your postings.

For example, suppose you posted a message about, say, painting, from your email address huckfinn@sawyer.com. You could do a simple search for *+painting huckfinn* to get all messages with painting, giving preference to those with your email address anywhere in them. Or, you could be more specific and search for *+painting +huckfinn@sawyer.com*.

Filtering Postings

You can also use AltaVista to filter messages for you. Suppose you subscribe to a newsgroup and—for whatever reason—you don't want to read postings from a particular author. News readers often have a feature called a Bozo filter or a Kill file to automatically screen out all messages from anyone you specify. You can do the same thing with your AltaVista searches. Say that robbie2957832@aol.com posts a lot of, in your opinion, irrelevant and misleading messages on your favorite topic of underwater macramé. You can set up a Simple Search for *"underwater macrame" -from:robbie2957832@aol.com* to get all the messages on your hobby, except those from this particular person.

Keeping Your Newsgroup Postings from Being Found

You can easily prevent your posting from being indexed by including the *X-No-Archive: Yes* command in these places:

- Anywhere in the header
- Anywhere in the first few lines of the message body
- Anywhere on the last line of the message

Your best bet is to include this command in the document header. Although AltaVista accepts this command when placed in the first few lines or last line of the message body, other search services may overlook the command in these places and index the document.

Tip *Use your best judgment when posting newsgroup messages. If you really don't want everyone in the world to have access to it, don't post it. Postings, particularly the ones you regret as soon as you send them, tend to have a life of their own and might end up where you don't want them.*

6

SOME POINTERS TO OTHER USEFUL CHAPTERS

Whew! Neat stuff! After you've mastered Usenet searches, you might visit one of these chapters:

- Hit Chapter 5 for a refresher course on Advanced Search.
- Visit Chapter 4 and learn how to use AltaVista's terrific Refine feature.
- Breeze through Chapter 9 for examples of how other people use AltaVista to create effective searches.

7

Customizing AltaVista

So far in this book, we've shown you how to use AltaVista's awesome power to quickly find the information you need. Now we'll show you how to customize AltaVista so that it's even easier to use. Imagine that!

In this chapter, we'll show you several different ways to customize AltaVista to meet your needs. In the first part, you'll learn different ways of making AltaVista even easier to access (and you thought it was easy already). You'll see how to make AltaVista your browser home page, set bookmarks, and even incorporate the AltaVista search form into your own Web site. Then, you'll learn to set AltaVista preferences, which are options you have for setting languages, designating defaults, and specifying the results format.

ACCESSING ALTAVISTA THE (REALLY!) EASY WAY

Although you can easily access AltaVista just by typing the URL (**http://altavista.digital.com**, in case you forgot) into your browser, you can make accessing it even easier by making some minor adjustments to your browser setup. For example, you can make AltaVista your home page in the browser, bookmark the AltaVista page for faster access to the site as a whole, or even bookmark results pages to be able to easily recreate searches. Let's take a look at each option.

Making AltaVista Your Home Page

One of the easiest ways to make AltaVista more accessible is to make it your home page. In doing so, you need only click the Home button in any browser to go right to AltaVista. Here's the process:

Note *The exact procedure will depend on the browser, the operating system, and the particular version of the browser you're using; however, these steps will get you pretty close.*

In Internet Explorer, use the following steps:

1. Go to the main AltaVista page by entering **altavista.digital.com** in the Address line of your browser.

2. After the page comes up, choose View | Internet Options.

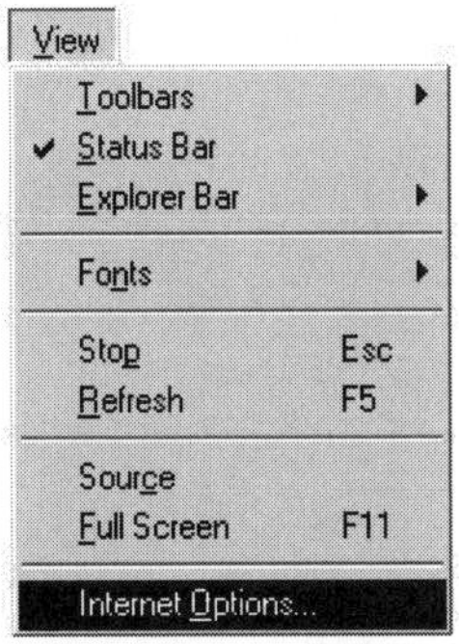

3. In the Internet Options dialog box, select the General tab.

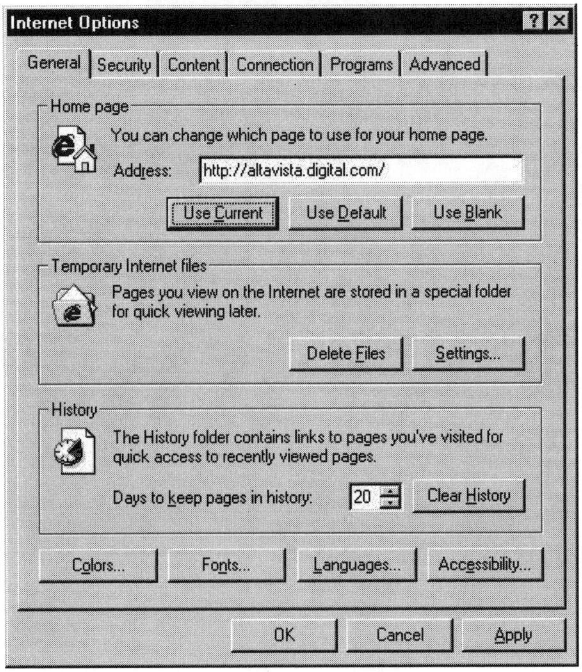

4. Click the Use Current button in the Home page section.

5. Click OK.

In Netscape Navigator, use the following steps:

1. Go to the main AltaVista page by entering **altavista.digital.com** in the Address line of your browser.
2. After the page comes up, choose Edit | Preferences.

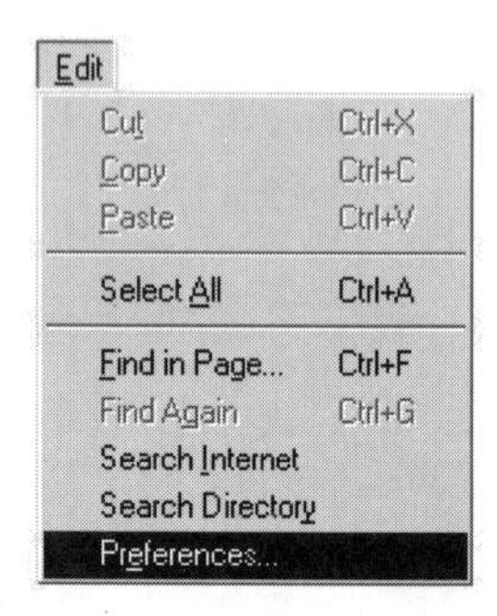

3. Select Navigator from the Category list on the left.

Preferences
Category:
Appearance
Fonts
Colors
Navigator
Languages
Applications
Mail & Groups
Composer
Offline
Advanced
Navigator Specify the home page location
Navigator starts with
Blank page
Home page
Last page visited
Home page
Clicking the Home button will take you to this page.
Location: http://altavista.digital.com/
Use Current Page Browse...
History
History is a list of the pages you have previously visited.
Pages in history expire after: 9 days Clear History
OK Cancel Help

4. Click the Use Current Page button in the Home page section.
5. Click OK.

Now, whenever you open up your browser, AltaVista will appear as your home page.

Bookmarking the AltaVista Home Page

Another option you have is bookmarking the AltaVista home page (or adding the page to your Favorites, depending on the browser you prefer), which lets you access AltaVista with just a couple of mouse clicks. For either Netscape Navigator or Internet Explorer, first go to the AltaVista home page, then follow these steps.

1. In Netscape Navigator, choose the Bookmarks button on the Location toolbar, then click Add Bookmark. To go back to the page, click Bookmarks (the first click), then just click on the AltaVista item in the list (the second click), as shown in Figure 7-1.

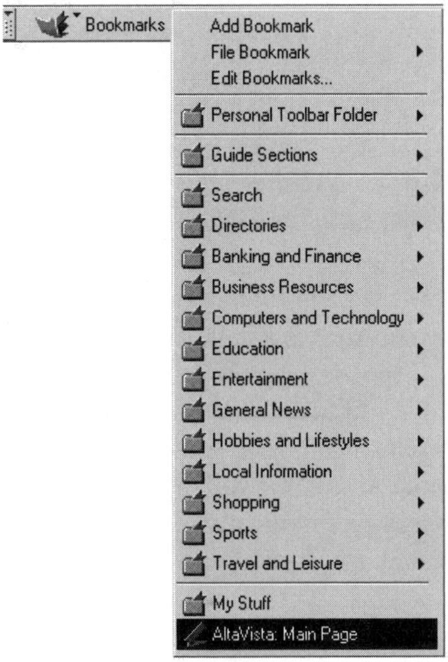

Adding AltaVista to your Netscape Navigator bookmark list makes it easier to return to AltaVista

FIGURE 7-1

2. In Internet Explorer, choose the Favorites item from the menu bar (at the top of the window), then choose Add to Favorites. After you add AltaVista to your IE Favorites list, it will be easier for you to find the main page.

3. Choose not to subscribe to the page, edit the name if you want, then click OK.

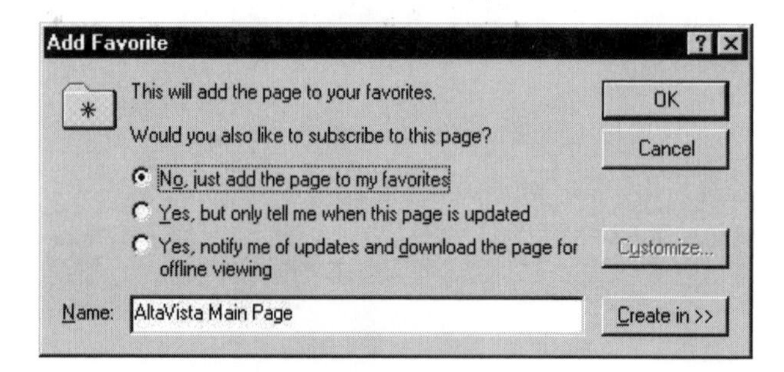

4. To go back to the page, click Favorites from the menu bar (the first click), then choose the AltaVista item from the list (the second click).

Pretty slick, huh? Additionally, if you conduct a search and you're particularly pleased with the results—and think you'll want to do the same search again—you could also bookmark the search. (We'd recommend doing this, by the way.) If you do that, your browser saves the search settings and information, but not the actual links. Each time you return to your bookmark, the browser resubmits the search to AltaVista Search and you get the latest and greatest list of links.

On either browser, go ahead and search as you usually would, then use the procedure just explained to bookmark the search. This works best if you've really come up with a good search, or you have a particularly complex search—there's not much point in doing it with a simple one- or two-word search.

Refer to your browser's help menu for more information about bookmarking and managing your list of Favorites. Of course, you could always just use AltaVista Search to find the help. Try searching for *managing bookmarks netscape* or *managing favorites explorer.*

Tip *Always use descriptive names for Favorites; otherwise, they'll all blend together and you'll have a bunch of "AltaVista Main Page" bookmarks, which won't help you out at all.*

Saving a Search

If you've just conducted a great search and want to save your results, you have another option. You could save the actual list of links (in the groups of ten that AltaVista Search presents). This choice would be good for putting together a customized list of HTML documents that you find useful. You'd be able to incorporate any or all of the pages generated by AltaVista Search into your document.

For example, if you're a librarian and want to help one of your clients get the most out of the search they just conducted, help them save the search—it'll make it much easier to find the pages again and to add them to a bibliography. Or, if you are helping your company get found more easily on the Internet, you could save searches to show your boss the progress you're making over time.

To save the results of your search in Netscape Navigator or Internet Explorer (and in most other browsers as well), do the following:

1. Conduct the search in AltaVista Search.
2. Go to File | Save As.
3. Select a filename and location for your document and click OK.

7

BUILDING ALTAVISTA INTO YOUR OWN SITE

In addition to making AltaVista your home page and adding browser bookmarks, you have another, somewhat more complex, option for easily accessing AltaVista—adding the AltaVista search form to your own Web site. AltaVista welcomes you to

use the following code in your personal home page; however, you'll need to get permission to use it in commercial sites.

For a pretty minimal AltaVista Search form with just a few options and bells and whistles, you can add the following lines of HTML code to your personal home page:

```
<FORM name=mfrm method=GET action="http://altavista.digital.com/cgi-bin/query">
<INPUT TYPE=hidden NAME=pg VALUE=q>
<INPUT TYPE=hidden NAME=text VALUE=yes>
<B>Search <SELECT NAME=what>
<OPTION VALUE=web SELECTED>the Web
<OPTION VALUE=news >Usenet</SELECT>
for documents in <SELECT NAME=kl><OPTION VALUE=XX SELECTED>any language
<OPTION VALUE=zh >Chinese
<OPTION VALUE=cz >Czech
<OPTION VALUE=da >Danish
<OPTION VALUE=nl >Dutch
<OPTION VALUE=en >English
<OPTION VALUE=et >Estonian
<OPTION VALUE=fi >Finnish
<OPTION VALUE=fr >French
<OPTION VALUE=de >German
<OPTION VALUE=el >Greek
<OPTION VALUE=he >Hebrew
<OPTION VALUE=hu >Hungarian
<OPTION VALUE=is >Icelandic
<OPTION VALUE=it >Italian
<OPTION VALUE=ja >Japanese
<OPTION VALUE=ko >Korean
<OPTION VALUE=lv >Latvian
<OPTION VALUE=lt >Lithuanian
<OPTION VALUE=no >Norwegian
<OPTION VALUE=pl >Polish
<OPTION VALUE=pt >Portuguese
<OPTION VALUE=ro >Romanian
<OPTION VALUE=ru >Russian
<OPTION VALUE=es >Spanish
<OPTION VALUE=sv >Swedish
</SELECT></B><br><INPUT NAME=q size=50 maxlength=800 wrap=virtual
VALUE=""><br><INPUT TYPE=radio NAME=act VALUE=search CHECKED>search <INPUT
TYPE=radio NAME=act VALUE=refine>refine <INPUT TYPE=submit VALUE=Submit>
```

You can delete any of the Option lines that specify languages you don't speak, or even delete everything from the <SELECT> tag before the languages start, through (and including) the </SELECT> tag after the list of languages to always get Any Language for the results. After you've entered this code in your home page, remember to save your document, then try it out it in a browser.

Alternatively, you could always start out with the whole code used to make the AltaVista home page and edit it down to size (assuming you're fairly familiar with HTML). Do remember to add

```
http://altavista.digital.com/
```

before each occurrence of

```
/cgi-bin/query
```

so you end up with

```
ACTION="http://altavista.digital.com/cgi-bin/query"
```

The best technique to use if you start with the whole AltaVista page and edit it down to size is the same technique that Michelangelo used for his sculptures—figure out what you want to end up with and take off anything that doesn't look like that.

SETTING ALTAVISTA PREFERENCES

AltaVista comes with a slew of preferences you can use to customize it to better meet your needs. Preferences give you the opportunity to be bossy and tell AltaVista what to do! In particular, you can do the following:

- *Tell AltaVista to search for pages published in languages you specify.* For example, by default, AltaVista returns matches published in any language. With this feature, you can specify that AltaVista return only matches published in languages you specify.

- *Choose AltaVista default views.* For example, you can specify that AltaVista's Advanced Search page be the default view, or you can specify that Refine's Graph View be the default.
- *Specify how much detail AltaVista provides with Web and Usenet results.* Remember, by default, AltaVista returns Web results in Detailed format and Usenet results in Compact format. Using preferences, you can specify Compact format for Web results and Detailed format for Usenet results, depending on your individual needs.

For a look at accessing the Preferences page and setting your preferences, read on.

Accessing the Preferences Page

Whether you're using Simple or Advanced Search, you can easily access the Preferences page by clicking the Preferences link, located below the search window. Figure 7-2 point outs the Preferences link on the Simple Search page. The link is similarly located on the Advanced Search page.

All you have to do is click Preferences, and you'll be whisked away to the Preferences page, as shown in Figure 7-3.

You'll notice that the Preferences screen is divided into three main parts: online instructions, languages, and a list of miscellaneous preferences. To set your preferences, follow these quick steps:

1. From the Preferences page, select your preferred options. For example, choose one or many checkboxes in the Languages section to specify that AltaVista return matches published in those languages. Alternatively or in addition, you can select from the checkboxes below the Languages section to specify default settings and views.

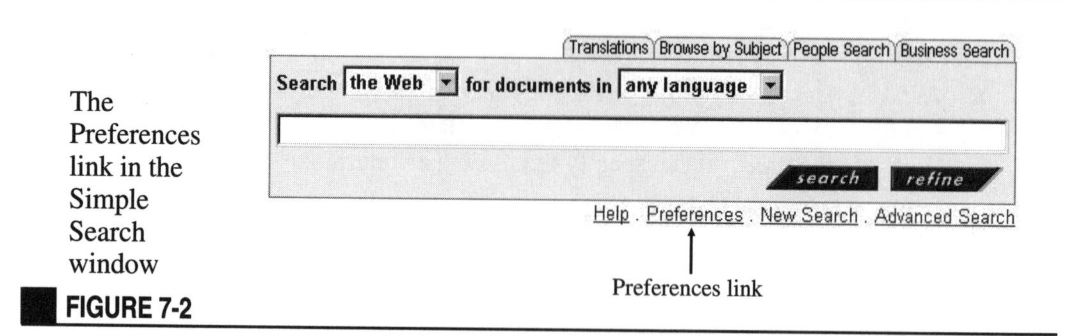

The Preferences link in the Simple Search window

FIGURE 7-2

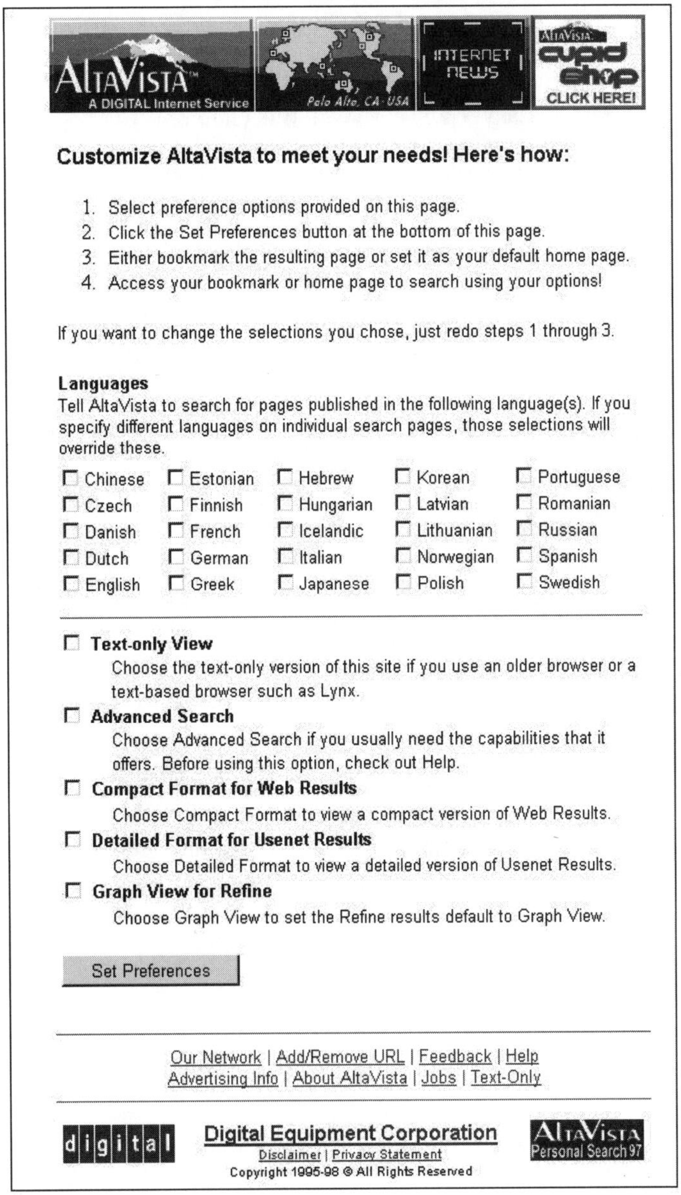

AltaVista™ A DIGITAL Internet Service | Palo Alto, CA USA | INTERNET NEWS | AltaVista cupid shop CLICK HERE!

Customize AltaVista to meet your needs! Here's how:

1. Select preference options provided on this page.
2. Click the Set Preferences button at the bottom of this page.
3. Either bookmark the resulting page or set it as your default home page.
4. Access your bookmark or home page to search using your options!

If you want to change the selections you chose, just redo steps 1 through 3.

Languages
Tell AltaVista to search for pages published in the following language(s). If you specify different languages on individual search pages, those selections will override these.

☐ Chinese	☐ Estonian	☐ Hebrew	☐ Korean	☐ Portuguese
☐ Czech	☐ Finnish	☐ Hungarian	☐ Latvian	☐ Romanian
☐ Danish	☐ French	☐ Icelandic	☐ Lithuanian	☐ Russian
☐ Dutch	☐ German	☐ Italian	☐ Norwegian	☐ Spanish
☐ English	☐ Greek	☐ Japanese	☐ Polish	☐ Swedish

☐ **Text-only View**
Choose the text-only version of this site if you use an older browser or a text-based browser such as Lynx.

☐ **Advanced Search**
Choose Advanced Search if you usually need the capabilities that it offers. Before using this option, check out Help.

☐ **Compact Format for Web Results**
Choose Compact Format to view a compact version of Web Results.

☐ **Detailed Format for Usenet Results**
Choose Detailed Format to view a detailed version of Usenet Results.

☐ **Graph View for Refine**
Choose Graph View to set the Refine results default to Graph View.

Set Preferences

Our Network | Add/Remove URL | Feedback | Help
Advertising Info | About AltaVista | Jobs | Text-Only

digital — Digital Equipment Corporation
Disclaimer | Privacy Statement

AltaVista Personal Search 97

The Preferences page offers all the customization options you could ever want

FIGURE 7-3

2. Click the Set Preferences button at the bottom of the page.
3. Either bookmark the resulting page or set it as your default home page. For a quick reminder, see the section called "Accessing AltaVista the (Really!) Easy Way" earlier in this chapter.
4. Access your bookmark or home page to search using your options!

In order for your preferences to take effect, you must remember to either access the bookmarked page or the home page in which you've designated your preferences. If you decide you don't want to use your Preferences for a while, just close your browser, restart it, and go back to AltaVista.

SOME POINTERS TO OTHER USEFUL CHAPTERS

Congratulations! You've just mastered some of the more detailed aspects of using AltaVista effectively. In fact, if you followed the steps presented in this chapter, you've made AltaVista your home page, set useful bookmarks, set AltaVista preferences, and even incorporated the AltaVista search form into your own Web site.

Where to from here? Check out these chapters:

- Breeze through Chapter 9 for search ideas and samples.
- Go to Chapter 10 to learn how AltaVista was developed and how it has evolved.
- Find interesting search trivia in the Appendices, such as the top 1000 words people search for, a sampling of queries, and the frequency of words indexed by AltaVista.

Providing Information the AltaVista Way

If you are one of the millions of people who have created (or will create) a Web site or who participate in newsgroups, this chapter is for you. As a content provider, you likely have a different perspective than you would if you were seeking information; you're likely to focus on having the content you provide be found by people searching the Internet, rather than on the act of finding information. Your biggest concerns are likely to be making sure that your potential readers can easily find information, ensuring that your pages appear as high as possible on results lists (of appropriate searches, of course), and encouraging readers to actually follow the link to your site.

While AltaVista will eventually find virtually everything on the Internet, as a content provider you can control potential readers' initial impressions of your site and help ensure that your site or newsgroup message makes it into your potential readers' results lists. You can tailor your information to take full advantage of AltaVista Search's unique capabilities, tell AltaVista as soon as your site is ready, and make sure that information is (or is not) found and indexed for the rest of the world to access.

That's what this chapter is all about—helping you provide information so that it's easily accessed by potential readers. In this chapter, you'll find everything you need to know about providing information effectively, including:

- An explanation of how AltaVista Search finds sites and how to submit your site manually, instead of just waiting for AltaVista Search (Scooter) to find you
- A thorough description of AltaVista Search's indexing techniques and ways to customize your documents
- Complete instructions for submitting your site to AltaVista
- Guidelines for not contributing to Internet index spam (improperly submitting pages and sites to search engines)

- Complete instructions on excluding your site or pages from AltaVista (and many other search services)
- Specific document design and structure tips
- A description of how AltaVista Search acquires news articles and how you can control the way your postings are indexed (or not indexed) by AltaVista Search

UNDERSTANDING HOW ALTAVISTA SEARCH WORKS FOR WEB PAGES

The Web—comprised of hundreds of millions of separate documents—is dispersed on hundreds of thousands of computers around the world. Capturing information from these millions of documents requires a computer program to visit each site and each page, just as a regular human user would, and retrieve all the text that is found there. These programs that automatically visit Web sites and gather information are known as "spiders" or "robots."

AltaVista Search's spider, called *Scooter*, sends out a thousand simultaneous connections to a thousand different Web addresses—also called URLs (uniform resource locators). It goes to each of those sites, submits a URL, and retrieves the complete text of the Web page. It then parses the pages and identifies other URLs mentioned, checks to see if they have already been retrieved, then goes and gets the new ones. That process continues until Scooter has reached all pages that are connected to other pages on the Internet. This technique enables Scooter to retrieve as many as ten million Web pages per day.

Note *Wouldn't all the commotion caused by Scooter retrieving pages clog up a Web site? Actually, no. To avoid such problems, Scooter pauses between retrievals from a specific site. The exact length of the pause depends on the response time of the site—when retrieving text from a slower site, Scooter will wait longer. This approach guarantees that if a particular machine is overloaded or has a slow Internet connection, Scooter will wait that much longer before returning for the second or third time.*

How Fresh Is the AltaVista Index?

Good question, and we're glad you asked. The AltaVista index provides extremely fresh information. Every six weeks or so, Scooter goes out and crawls the entire Web anew—the index produced through this new crawl completely replaces the old index. Additionally, the most popular sites on the Web (in addition to governmental sites and other sites of particular value or interest) are crawled continuously. Thus the oldest links in the entire AltaVista index will be only about six weeks old, the most popular links will be only a day or so old, and there won't be antique non-functional links for users to contend with.

And, in addition to Scooter's regular roaming, a Scooter sibling, Microscooter, retrieves Web pages, dashing out and retrieving the pages that Web content providers submit to AltaVista (through the Add URL link). Those pages are crawled immediately and indexed within a day or so, and the pages they link to are added to Scooter's to-do list for regular crawls.

Now, that's fresh!

As remarkable as Scooter is, it does not retrieve quite everything. For example, there are some pockets of pages that link to one another but that have no connections with the rest of the Web—these pieces of information might not be found. Additionally, there are entire private networks behind firewalls, including most corporate networks, that Scooter cannot access. Pages that don't include plain text, like graphic and multimedia-focused pages, won't be effectively retrieved. Finally, Scooter does not retrieve those pages where Web masters have indicated that they do not want to be visited by robot programs (using the Robot Exclusion Standard, described later).

A good way to understand Scooter is to think of someone using a text-only browser or of a blind person using a text-to-speech synthesizer searching the Web. Any text that's available will be understood, but images, multimedia, and frames, not to mention video and sound files, will go unseen and unindexed.

How AltaVista Indexes Sites

After Scooter retrieves a site, AltaVista builds an index of every word in every document using custom-designed indexing software. By taking this approach rather than developing a directory or structured database, AltaVista Search deals with chaotic information in its natural disorganized state and doesn't try to turn the Internet into something it isn't—like a telephone book or an encyclopedia. AltaVista Search lets you find a needle in a haystack without having to rearrange the haystack first.

Note *The same basic indexing software is used for both newsgroups and the Web, though the indices themselves are separate. Both Simple and Advanced Search use exactly the same index and the same indexing software.*

How AltaVista Search Ranks Sites

When you submit a Simple Search query, the results you get are a listing of ranked documents, not just a random listing of documents AltaVista Search found on the topic you submitted. How does AltaVista know which matches to list first, second, and so on? AltaVista Search ranks the information you receive using two elements:

- The rarity of words you provide in the search query
- The location of search words in the documents

The most important ranking factor is the rarity of words you provide in the search field. When a potential reader submits a search, AltaVista ranks the results based in part on how uncommon the terms are. Extremely common words (e.g., *a, and, the, Internet, computer*) are ignored for purposes of ranking because they appear in so many documents that they provide no help in establishing how relevant a given document is to your query. That's why it's important to enter rare words in your search queries, as we mentioned in Chapter 2.

Note *On the results page, along with the word count, you see which words were "ignored."*

The next most important factor for ranking is the position within the document of the query words and phrases. AltaVista Search presumes that position within a document often indicates the relative importance of a word to the document as a whole. For example, a word in a title is generally more indicative of the content of

the document as a whole than a word in the body of the text several screens deep in the document.

Specifically, AltaVista Search prioritizes according to the following rules:

- Pages that have the target words in the HTML title and near the top of the text are ranked higher than others. This rule includes any content in META tags in the head of the document.
- Pages in which the words appear near one another in the text are ranked higher than those in which the words are farther apart.
- Pages in which the word appears twice are ranked higher than those with only one occurrence. (More than twice makes no difference.)

Note *The precise algorithm used to rank documents is very complex and includes many more factors than those mentioned here, but these are the key factors.*

The indexing software looks through the documents in the index and considers every document that includes at least one of the words in the query, calculates the score for each matching document, ranks the matching documents according to score, and then generates a prioritized list of Web pages.

Note *Because of the immense scope of the information indexed—tens of billions of words in tens of millions of documents—AltaVista Search uses approximations for ranking. Rather than determining precisely if a given word appears in 149,985; 150,003; or 150,010 different documents, AltaVista Search rounds off to 150,000—a good approximation.*

What AltaVista Search Does Not Index

As an information provider, you must ensure that AltaVista can index enough of your site that your potential readers can find it. Normally, AltaVista indexes only HTML or text formats, which seems fairly limited since many Web sites contain so much more than these basic formats. And, although AltaVista indexes a heck of a lot of pages in millions of sites, it does not reach every element of every page. In particular, AltaVista does not index

- Pages it can't reach because they require registration or password-based access

- Text within images
- Text in non-HTML or non-text files, such as Acrobat or Postscript files
- Multimedia files
- Extremely long (over 4MB) text files
- Some dynamic content
- Frame context

We'll take a closer look at these in the next several sections.

Registration or Password-Based Access

Some sites require registration or passwords for users to access the information. For example, some sites welcome visitors to the home page and then require them to go through some additional step, such as filling out a form, registering, or providing a password before proceeding further. Procedures like these stop Scooter in its tracks. If that's how you've established your site, only your home page will be indexed. That means that people checking AltaVista Search to get an idea of where to go for your kind of information or service may not even realize that you exist.

Heavy Use of Images

AltaVista cannot read the text within images (such as in GIF and JPEG files). That means that although your page may have lots of useful content *for people*, Scooter can't read any of it. Be sure to use META tags to help supplement your pages with text, if you rely particularly heavily on images. And, take care to use ALT text for your images to provide additional content that Scooter can feed to the index.

While AltaVista cannot read images, it can read a client-side image map. If you have pages linked through a client-side image map, Scooter will find them. However, if you use an old-style (server-side) image map, AltaVista may miss some of your pages.

Just think of Scooter as a browser without the ability to view images and you'll be fine. If readers can use your site and get the information they need without being able to see images, Scooter will be able to handle it.

Specialized Formats

Many sites use specialized formats rather than HTML or text-based formats to present information. For example, Acrobat files are attractive and usable for many people, but they do not readily lend themselves to full-text indexing by AltaVista or

other search engines. Although these files do contain text, AltaVista cannot read them, and people using AltaVista will not know that you have such rich content—all they might find is the filename. You should at the very least make your filenames clear and descriptive, or, far better, provide a plain-text version as a supplement to these fancier files.

Multimedia Files

More and more, people are adding nifty effects to their Web sites, such as audio and video files, Shockwave format files, Java applets, or RealAudio or RealVideo files (collectively known as multimedia files). While multimedia files often make your pages more interesting, keep in mind that only their filenames and the ALT text associated with them are indexed by AltaVista. If you choose to include multimedia files, be sure to give them descriptive names, provide detailed alternative text, and (if possible) provide alternative forms of the same information.

Large Files

Enormous text files (over 4MB, or 2,000 pages of text) are also not completely indexed by AltaVista. Documents larger than 4MB are truncated to 4MB by Scooter when it fetches them. Then, in the current index, documents larger than 100K have everything in their first 100K indexed and only links indexed thereafter. So if you have text files larger than 100K and you want their full contents indexed, you should consider breaking them into a series of smaller files, by part or chapter. (This step will also make the files *much* easier for your readers to use.)

Dynamic Information

If your pages change very often—either to suit the profiles of individual users or to stay constantly up-to-date with the latest news, like stock quotes—keep in mind that AltaVista won't index the dynamic information. When AltaVista encounters a dynamically changing site, it indexes the static content—for instance, the front page, help, and background pages—and whatever it happens to find the day it visits.

Additionally, some dynamic content cannot be crawled. Generally, problems occur when content is automatically generated from a database or server-based scripts. There's no reliable way to predict exactly what will or won't work—the best solution is just to try submitting the page (as described later). If AltaVista has problems with it, you'll get an error message announcing that the URL was invalid. After you've submitted the URL of a dynamically generated page, also check back in a couple of days and make sure it's there—if not, send an email message to

suggestions.altavista@pa.dec.com, mention the URL you submitted, and ask if it will be successfully crawled by AltaVista.

Additionally, pages that use "cookies" (to identify visitors when they return) in the URL are also eliminated from the index, simply because the text indexed would not match what a user would see when linking to that site through AltaVista. See "Excluding Pages or Sites from AltaVista" later in this chapter for more information about automatically keeping these pages from AltaVista Search.

Documents in Frames

AltaVista understands frames to a certain extent. Scooter will follow frame links and index individual framed pages. However, AltaVista cannot preserve the context of the pages within the frames.

What's the context? Frames have a master page (known as a frameset document) that tells your browser how to place the initial framed pages within the frames. Since AltaVista indexes the framed pages directly, visitors will be taken directly to those pages. Visitors through AltaVista will quite likely never encounter the master page, and so they miss the context the framed page should be presented in. Additionally, often pages are linked only within the context of frames, and never appear directly. There is often no single URL that points to a particular configuration of content within frames (the configuration might, for example, result from two links from an original frame configuration).

A variety of solutions to this problem exist, but they go beyond what can be described here. See the Help pages on the AltaVista site or Search Engine Watch's "Search Engines and Frames" tutorial at **http://searchenginewatch.com/webmasters/frames.htm** for more help on this subject.

SUBMITTING YOUR SITE TO ALTAVISTA

Normally, there is no need to tell AltaVista Search that you exist, that you have changed your pages, or that one of your pages has gone away. Remember, Scooter indexes 10 million pages daily and does a full-Web search once every six weeks. However, there are a few circumstances in which you would want to submit your URL directly to AltaVista Search:

- Your site is relatively new and very few or no other sites have pointers to it.
- Your site is already in the AltaVista Search index, but you recently added new pages or made significant changes to existing pages.
- You want to make your site accessible as soon as feasibly possible.

To submit your site, just follow these quick steps:

1. Click Add/Remove URL, located at the bottom of most AltaVista pages. You'll see a screen similar to Figure 8-1.
2. Read the instructions and information for the latest submission policies.
3. Enter the URL for your Web site in the URL field.

That's it! AltaVista retrieves the page right away, indexes it quickly, adds the information to the index, and then adds your URL to the list of sites for Scooter to visit as soon as possible. By following the links from your home page and the subsequent links from there, Scooter will be able to find the rest of your pages.

Note *You can submit more than one page to Scooter. You may want to do this if you add a series of important pages that should be indexed right away. However, you don't want to submit more than five or so pages on any given day. Many index spammers submit hundreds of pages, so Scooter has an unpublished limit on how many it will accept. It varies by Web site. If you stay in the range of five pages, you should have no problems.*

Some people using popular service providers or Web hosting services, such as America Online or GeoCities, may get an error message announcing that Scooter has reached its limit of submissions for that day. If this happens to you, try back at another time, probably earlier in the day. Many people submit pages from these addresses, so Scooter's limits are higher than with other sites, but the limits can still easily be reached on a given day.

Note *Scooter will also revisit existing sites and notice when pages no longer exist. They will automatically be purged from the index.*

If you need to remove a page from the index—for example, a page that doesn't exist any more, just submit the old (and invalid) URL. As soon as AltaVista finds that the site returns a 404 (Site Not Found) error, the URL and associated information will be removed from the index. If you need to remove a page that still exists from the index (for example, if you want people to just enter your site through the front door and not through other pages), you'll need to create a ROBOTS.TXT file (as

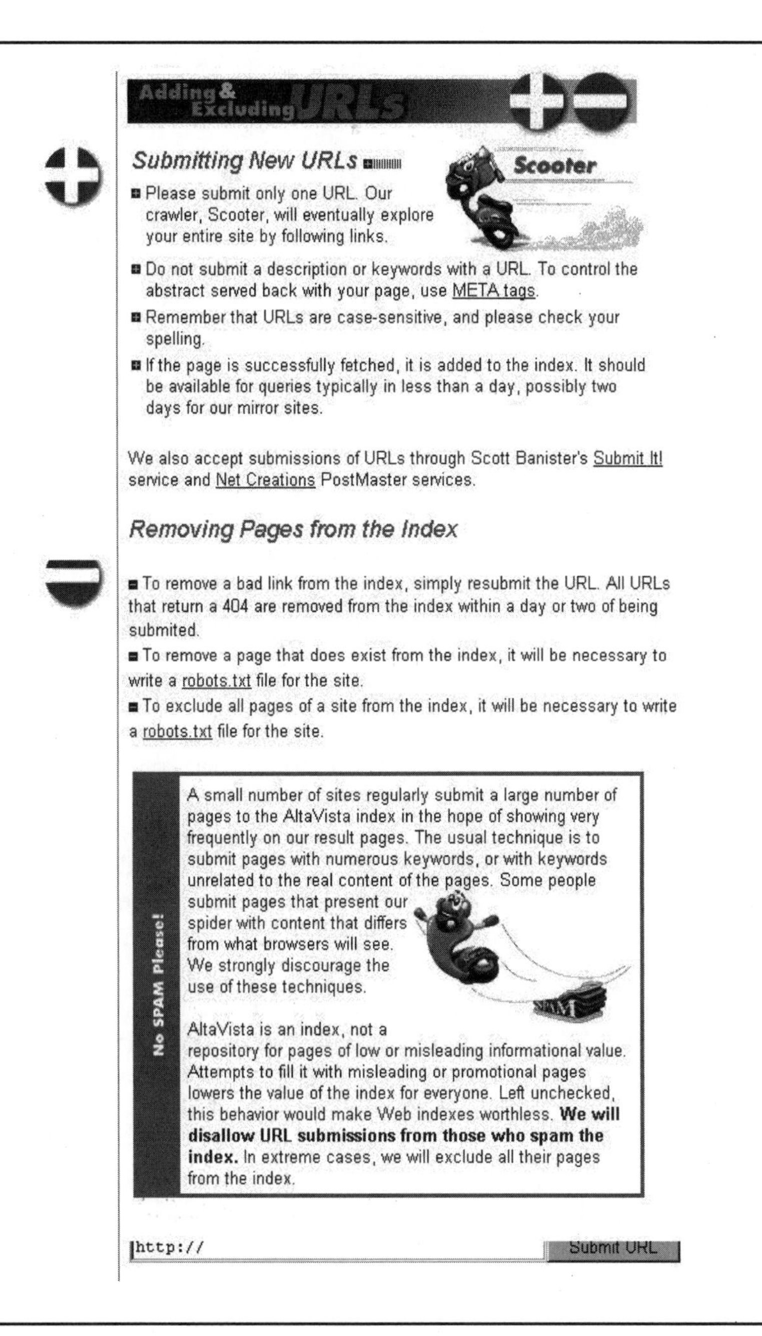
Adding & Excluding URLs
Submitting New URLs
Scooter
Please submit only one URL. Our crawler, Scooter, will eventually explore your entire site by following links.
Do not submit a description or keywords with a URL. To control the abstract served back with your page, use META tags.
Remember that URLs are case-sensitive, and please check your spelling.
If the page is successfully fetched, it is added to the index. It should be available for queries typically in less than a day, possibly two days for our mirror sites.
We also accept submissions of URLs through Scott Banister's Submit It! service and Net Creations PostMaster services.
Removing Pages from the Index
To remove a bad link from the index, simply resubmit the URL. All URLs that return a 404 are removed from the index within a day or two of being submited.
To remove a page that does exist from the index, it will be necessary to write a robots.txt file for the site.
To exclude all pages of a site from the index, it will be necessary to write a robots.txt file for the site.
No SPAM Please!
A small number of sites regularly submit a large number of pages to the AltaVista index in the hope of showing very frequently on our result pages. The usual technique is to submit pages with numerous keywords, or with keywords unrelated to the real content of the pages. Some people submit pages that present our spider with content that differs from what browsers will see. We strongly discourage the use of these techniques.
AltaVista is an index, not a repository for pages of low or misleading informational value. Attempts to fill it with misleading or promotional pages lowers the value of the index for everyone. Left unchecked, this behavior would make Web indexes worthless. We will disallow URL submissions from those who spam the index. In extreme cases, we will exclude all their pages from the index.
SPAM
http://
Submit URL

FIGURE 8-1 The Add/Remove URL screen

described later in this chapter) that excludes the page from being indexed, then resubmit the page and site to AltaVista.

Avoiding Index Spamming

As an information provider, you face a significant challenge: getting your site or posting found. Rather than having just a few thousand Web sites to compete with, as was the case a couple of years ago, you now have over a million—and counting—to compete with. Your challenge is to ensure that your site appears within the top few search results for a given query, keeping in mind that getting your site listed within the first several matches is often the difference between people visiting the site or not.

This challenge to get pages listed high within a results list has brought about a new problem, called *index spamming*, which is an attempt by spammers to manipulate search services like AltaVista to

- Force pages to appear above similar pages in a results list
- Force many pages into the index (making it more likely that their pages would be found)
- Deceptively attract additional visitors that would not intentionally have visited the site

Of course, these manipulations, if successful, would reduce the overall quality of the index and diminish the value of the search service for everyone. How? Well, spamming skews results, obscures relevant information, and fills results lists with information unrelated to the search query. The result is that information seekers often find several links that lead to one site or find links that lead to completely unrelated information. For information providers, index spam can make it harder to get found and indexed, making a site less accessible. For AltaVista as well as other search service providers, successful index spam reduces the overall index quality and diminishes the value of the service for everyone.

How much spam does AltaVista receive? A lot. On a typical day, submissions through AltaVista's Add URL page might total somewhat more than 20,000 URLs. Of those, nearly 12,000 pages will be pure spam—intentional duplicates of other pages already indexed or spam of other varieties—and won't ever make it into the index. Only 8,000 new pages generally remain from the submissions.

Certainly some people will, in good faith, attempt to get their pages effectively indexed and inadvertently do something that's broadly considered index spamming.

And perhaps some people don't realize that the techniques they use are, quite simply, wrong. These two types of people do not create the major problems associated with index spamming. Usually, any issues are one-time occurrences with small sites and are easily addressed through a couple of educational email messages.

Systematic spamming, however, poses a much larger-scale problem. This category of spammer includes a whole sub-industry, in which search-engine placement services promise that for a few dollars a company can get top placement in search services. This type of spammer seriously skews search results and makes search services less effective for information seekers and providers alike.

Though the widespread significance of index spamming has only recently been realized, Digital Equipment Corporation has made solving the index spamming problem a top priority and has taken several steps to ensure AltaVista's search results are as accurate, complete, and useful as possible.

First, Digital's research labs—the same ones that developed AltaVista—have turned their attention to the problem of index spamming. Among other characteristics of index spam, they've developed detection and defense software that indicates with very high reliability which pages are likely to be spam.

- Pages flagged as definite spam (for example, dozens of identical pages at the same site, submitted at nearly the same time) are automatically excluded from the index.
- Pages that are possibly spam appear on reports, which are reviewed by Digital's AltaVista team members. Problematic sites are then visited and revisited to further examine them. Spammers that fall into this category are contacted and informed about index spam and given an opportunity to correct the problem. If spamming continues, these sites and submissions are blocked from the index permanently.

Second, Digital's AltaVista Service is developing guidelines to help ensure that search results are high quality and valid, and that they reflect the information that information seekers are looking for. Although these proposed policies may not totally prevent spamming, they will allow search-engine administrators to form a united front against identifiable index service abuses. These proposed policies, intended to guide Web content providers in getting found and indexed the right way, can be summarized as follows:

- Don't be deceptive or misleading about the content of your site.

- Use META tags correctly, including accurate information about your site.
- Be simple, straightforward, and accurate with descriptions and significant words.

In brief, anything in disguise as something else—like a life insurance page masquerading as a stock-purchasing tips page—is spam. Note that Digital's AltaVista does not judge the content or value of submissions; potentially objectionable content does not constitute index spam, but potentially objectionable content disguised as something innocuous does.

Excluding Pages or Sites from AltaVista

Because AltaVista Search searches and indexes the entire Web, it could eventually find any Web page that is connected to the main body of the Web through even one hypertext link. Even if no links lead directly to the page, if the URL is known and submitted by anyone to AltaVista Search, the information will be available to anyone on the Web. Because any information you have on a Web site could potentially be found by AltaVista Search, you might consider specifically excluding sensitive pages from AltaVista Search.

Why Exclude?

You might choose to exclude pages or entire sites for a number of reasons. First, you might want to exclude your entire site while it's still being constructed. Because your site represents you or your company, it's important that the content and structure are in place and accurate before you make it available to the world, particularly through an accidental release into the world's largest Internet search engine.

Second, you might want to exclude some pages from AltaVista Search to try to maintain some control over the context of the user's experience. You could allow pages that can serve as entry points to be indexed, and block other pages that might be confusing if accessed directly. For example, if you're running a contest with a three-step procedure to enter, you might want to block steps two and three from AltaVista Search so readers would have to start at the first step.

Third, you might want to exclude pages that only specific people should access. For example, you might exclude pages that are still under construction or that are being tested. You can provide the address to the testers or the people within your organization, but you wouldn't want them indexed on AltaVista Search.

Fourth, you might want to control which parts of your site are indexed and which are not. For instance, you may have pages that you would like to make available on the

Web for the convenience of those who need them, but would prefer not to publicize broadly. You might, for example, want to make usage logs available without making it particularly easy for everyone on the Internet to retrieve that information.

Note *Don't assume that if there is no hypertext link from any Web page to some test or "confidential" page of yours that no one can find it. Any documents on a public Web server are at least potentially available to everyone on the Web. In fact, some Web server software has a "directory indexing feature" that lets Scooter and other robots see the entire contents of a directory, and, therefore, find and pick up files that have no links to them—pages you might consider work in process or junk. If you have any files that you would prefer that the public not see but that you must keep on a Web site, you should check your server's configuration and make sure it isn't making that information available to the world.*

How to Exclude Pages or Sites

If you do have reason to exclude pages or sites, you can keep them out of AltaVista Search, and it's easier than you might think. AltaVista Search adheres to the Robot Exclusion Standard, which is basically an Internet honor system that specifies how robots and spiders are supposed to behave. This standard makes it easy for you to block spiders or robots from indexing your entire site, from finding particular directories, or from looking at particular files.

If you're sure you want to keep parts of your site from being indexed, simply set up a file named ROBOTS.TXT on your Web server in the main (top-level) directory of your Web server. This file should contain a list of the directories and files that you want to deny access to.

For instance, the file in the following example would exclude all spiders (*) from three directories (cgi-bin/sources, access_stats, and cafeteria/lunch_menus).

```
User-agent: *
Disallow: /cgi-bin/sources
Disallow: /access_stats
Disallow: /cafeteria/lunch_menus/
```

Any URL matching one of these patterns will be ignored by robots, like Scooter, that abide by the Standard. Of course, this file won't help at all if you get a disobedient robot, but that's life. So to speak.

Controlling How AltaVista Search Indexes Your Site

Even though AltaVista Search does not index all information from all pages, you can control how AltaVista Search indexes your site. AltaVista Search provides you with a few quick-fix alternatives that help you direct readers to your Web site. This section outlines ways that you can help potential readers find you through AltaVista Search as well as other tips that might make your pages more usable in general.

As an information provider, you have several tools in your arsenal to ensure that your potential readers can find your pages. There are several areas you can control that will have a significant effect on how AltaVista Search indexes your site:

- HTML document title
- Description
- Keywords
- Content
- Date

Note *There are no tricks that allow you, as Web master, to ensure that your pages appear on the top of results lists. The best you can do is to be clear, direct, accurate, and complete in all your descriptions, using the words that people who want to find you are most likely to use and understand.*

HTML Document Title

HTML document titles are the first thing your potential reader will notice in the list of search results. Titles are not the headings that appear in large, bold type at the top of Web pages, but rather, they are the text lodged between the TITLE tags in the HTML document source.

If you're counting on readers' finding you through AltaVista Search, you need to pay particular attention to the HTML titles for two reasons: First, the HTML title is usually the piece of information people use to decide whether to visit—or not to visit—your Web site. Because of this, you should make sure that you include it and that it clearly describes the page.

Second, the HTML title is important because words in the title are given priority for ranking purposes. If there are words that people would naturally search for when they want to find a site like yours, make sure those words are in the title. In this context, accuracy and completeness are extremely important.

HTML document titles appear in the source of a page between the tags <TITLE> and </TITLE>. If you're using a browser like Netscape Navigator or Internet Explorer, you will see the title of a page only at the very top of your browser window, in the title bar.

Note *Some content providers use existing pages as templates when creating new ones, and unwittingly forget to give the new page a new title. That is why results lists from AltaVista Search sometimes show several different entries with the same title, titles that don't seem to be accurate, or pages that don't have any title at all (listed most unattractively as "No Title").*

Description

The second element an AltaVista Search user is likely to notice in a results list is the abstract, or description. The abstract simply provides some data about what information your site contains. In many cases, AltaVista Search just uses the first few lines of your HTML document for the abstract. If those lines are (or could be) appropriately descriptive, great! However, if the first few lines are not particularly descriptive or if you want to present a more global description of your site, you can tell AltaVista Search exactly what you want the abstract to say.

To specify what you want the abstract to say, include the META tag between the document HEAD tags. For example, suppose you're the Web master for ACME Corporation and one of your pages starts like this:


```
<HTML>
<HEAD>
<TITLE>Developments at ACME Corporation Innovative Devices</TITLE>
</HEAD>
<BODY>
<H1>ACME Corporation Innovative Devices Information</H1>
```

You can add a META tag to the document head with a brief document abstract, using this format:

```
<META name="description" content="ACME Research and
Development efforts continue to be successful">
```

The top of your document would then look like this:

```
<HTML>
<HEAD>
<TITLE>Developments at ACME Corporation Innovative Devices</TITLE>
<META name="description" content="ACME Research and
Development efforts continue to be successful.">
</HEAD>
<BODY>
<H1>ACME Corporation Innovative Devices Information</H1>
```

After Scooter revisits your site, AltaVista Search will return your new description with the URL when people search for information contained at your site.

Keywords

Another aspect you should consider to help your pages be more accessible are keywords. *Keywords*—any particularly descriptive terms—augment the content of your HTML documents by adding other terms through which your potential readers could find you. Keywords are not directly visible to your potential readers, but they do help ensure that your potential readers will find your site when they need to.

The content alone of your document, taken out of the context of the rest of your site, might not necessarily be sufficient for potential readers to track you down. For example, as the Web master for ACME Corporation, you might have placed all your company's press releases online. Whenever people visit your site, they can select a link for the latest information about the company, and that's worked well so far. However, the phrase "press release" probably never appears on any of those pages. In other words, someone searching for *"ACME Company" AND "press release"* might find nothing.

There are two possible remedies for that kind of situation. First, you can simply edit the text to add the necessary words and phrases. Second, you can designate keywords that may or may not appear in the text by using the META tag. This section covers adding keywords using the META tag.

Adding keywords to a document simply requires editing the HTML code and adding information within the HTML HEAD tags, as in the following example. Before AltaVista Search, the top of an HTML document on your site might have looked like:

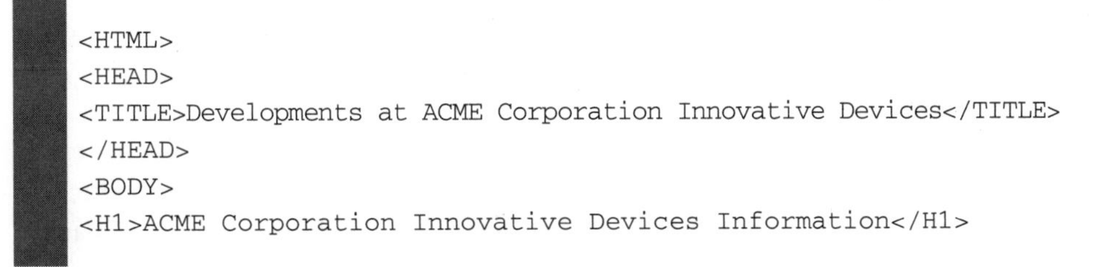

```
<HTML>
<HEAD>
<TITLE>Developments at ACME Corporation Innovative Devices</TITLE>
</HEAD>
<BODY>
<H1>ACME Corporation Innovative Devices Information</H1>
```

For instance, to indicate that a particular document is a press release, create a META tag with the following information:

```
<META name="keywords" content="press release">
```

If you need to add additional keywords, you can throw them in as well.

```
<META name="keywords" content="press release Wile Road Runner
failure still trying">
```

The top of your HTML document would then look like:

```
<HTML>
<HEAD>
<TITLE>Developments at ACME Corporation Innovative Devices</TITLE>
<META name="keywords" content="press release Wile Road Runner
failure still trying">
</HEAD>
<BODY>
<H1>ACME Corporation Innovative Devices Information</H1>
```

Your keywords can total up to 255 characters long. That includes spaces, and it counts everything that appears between *content="* and ">. If your description is

longer, AltaVista will just ignore the additional characters. Also, you do not need to use commas in your tag to separate words.

AltaVista Search will match any of the words listed, even though a visitor to that page would not see the META tag, and might not see those words anywhere in the text. Someone searching for *+"ACME" +release* would not have a problem finding you. Additionally, searches for *ACME road runner Wile* would now also get your page as a match.

When deciding which words are "key" for you, take advantage of unique terms associated with your company and its products. You should use typographically unique trademarks, model numbers, and particularly descriptive or unusual names whenever possible.

Note *There's no need to attempt to load your document with appropriate terms by using comments tags (<COMMENT></COMMENT> or <!-- -->) in the header or the body of the text. AltaVista Search deliberately does not index information in comment tags so your notes to yourself remain your own.*

As an additional example, if you are developing a site selling stereo equipment, you could add keywords like *high-fidelity*, *music*, *stereo*, *quality*, *component*, *speakers*, and the like to your documents. Consider any and all keywords and phrases that might reasonably occur to people looking for the kind of information you provide and add those words to your documents.

Dates

The third item that your potential readers will see within search results is the date that the page was last modified. This date reflects the last time that particular page was modified, not necessarily the date other information in the site was last modified. This date might be misleading for your potential readers because you might have, in fact, updated some pages in your site but not the particular page that AltaVista Search retrieved.

No matter how good your content may be, if people see a date like May 15, 1994, prominently displayed at AltaVista Search, they'll often presume the site is inactive and the information is obsolete. The date is also important to Scooter, which only looks for new material. If for any reason, the version number or date of a page has not changed, Scooter will not revisit it, and new material there will not be found. That means that the page will continue to be listed, but with the old content. Therefore, when updating pages in your site, you need to be sure that you also update all pages related to the one you're updating so that they all reflect a new revision date.

Additionally, you have to make sure that your server is properly configured to send out the correct date and time. AltaVista Search has found files dating from the first half of the 20th century, as well as some from the 21st. Presumably these file dates result from server error, rather than a time warp.

Designing Your Site for Best Results

Because users can come and go as they please, the structure of a Web site should be cohesive enough to hold readers' attention and to lead them through a controlled sequence of choices. Many—probably most—Web sites are organized hierarchically, with the home page as the top element, providing text links to the next level, which in turn provides links to the next level. Such a structure can make it easy for a user who comes in by the front door (or home page) to navigate smoothly around the Web site.

Before AltaVista Search, it was reasonable to assume that people navigating to and through a Web site would start with the home page and then follow through this hierarchy. But with AltaVista Search, every indexed page is equally likely to be found. AltaVista Search users are likely to bypass your home page and arrive directly at any of the pages deep in your carefully constructed directory structure. Therefore, you can no longer presume that the user has followed a certain predefined path to get to any one of your pages. So as an information provider, you should keep some additional concepts in mind when creating for this new environment:

- Provide logical and clearly labeled navigational links. Terms like "next page" and "previous page" aren't nearly as useful as links labeled by content, such as "Acme Corporation's Mission Statement."
- Clearly state the purpose of each page within the page itself. You cannot assume your reader will gain the context from other pages at your site—they might not even see other pages at your site.
- Provide links back to the main page of the site from each page in the site, in addition to information about the page maintainer and contact information.
- Look at the full text of your pages, not just the titles and descriptions. Remember, AltaVista Search indexes information from the entire page, not just from titles and descriptions.
- Make sure that each and every page can be understood by people who cannot see pictures. All graphics should have text alternatives for

AltaVista Search to find, and all words should be able to stand on their own, without graphics or multimedia files.

When you're creating Web pages, try to imagine the ways in which potential visitors to your site are likely to search. Varying your writing style by using synonyms rather than repeating the same word numerous times will not only make your text more readable, but will also increase the likelihood that people will find it.

Remember *You also can put an AltaVista Search query box right on your Web page, for the convenience of your users. The brief code that you need to do that is available at the AltaVista Search site. Simply save it and paste it into your Web page document. The simplest way to do that is to click on FAQ on the bottom of the AltaVista Search page. You'll see a section that provides code that you can just copy and use on your page. The result will be the AltaVista Search submission form, without the AltaVista Search graphics. People can enter their queries from the context of your site and get answers back from AltaVista Search.*

USEFUL TECHNIQUES FOR WEB MASTERS

One of the biggest challenges for a Web master is just keeping up with everything happening in a site. As pages and links come and go (often without notification or warning), keeping track of all the pages and how they relate can be an imposing task. The following tips and techniques will help you, as a Web master or information provider, use AltaVista Search to do your job more effectively.

Keeping Private Information Private

If you work for a corporation as a Web master, you are probably partially responsible for making sure that information that should not be public isn't available through your site. However, you also probably don't create all the pages on the site, nor do you have time to read all those pages carefully. Following is a solution that will help you keep private information from going public.

One of the first things you should do is conduct a query for documents at your site that contain any corporate information security markings. For instance, you might try the following:

```
host:yourdomain.com "company confidential" "proprietary" "top
secret" "internal use only"
```

Hopefully, a document with such warning labels wouldn't get posted on your site in the first place. But if and when such a mistake is made, you will want to know about it right away. Just bookmark that query and click on that bookmark regularly.

Similarly, as you learn of other sensitive projects and topics, create a query for those terms, bookmark the query, and use that bookmark regularly too. You might have a simple search query like the following:

```
url:yourdomain.com "Project X" "Customer X" "Acme Road Runner
Cannon"
```

The same approach applies to any sensitive information at your site.

Taking Inventory of Internal Links

Because your readers might enter your site at any page within the site, you should pay special attention to the hypertext links on all your pages, making sure that they provide good navigational clues and information about the rest of your site. You can use AltaVista Search to do a quick inventory of the internal links at your site. For instance, you can submit a query for links and URLs at your domain name, like this:

```
+link:yourdomainname +url:yourdomainname
```

The results of this example give you a list of pages with hypertext links that connect pages within your site that AltaVista has indexed.

This list of results will provide the information you need to see which pages link to other pages, and which do not. For example, if you find a number of pages that don't include a link to your home page, you might need to look into rewriting some of the pages so people who just find those pages can also find your home page.

Keeping Up with External Links

AltaVista Search also allows you to check which other Web sites and pages have hypertext links to your pages, and what particular pages they are linking to. Again, you can submit a query for links and URLs at your domain name, like this:

```
+link:yourdomainname -url:yourdomainname
```

The results of this query give you a list of all Web pages that AltaVista has indexed outside of your Web site that point to pages at your site. If there are less than 200 such pages, you can see them all with this Simple Search query by using the Next link at the bottom of the page to see successive screens of results.

Remember *If there are more than 200 pages and you want to capture information about lots of them, use the following line in Advanced Search (leaving the Ranking box empty):* link:yourdomainname *AND NOT* url:yourdomainname.

Check the context in which these pages point to you and whether they point to your home page or other specific pages at your site.

Tip *You can also use AltaVista Search to find other sites that are related to your business and might be of value to your audience but that are not competitors. You might then arrange for reciprocal links with the best of them, especially those that complement the material on your own pages.*

Fixing Broken Links

As a Web master or information provider, there are two kinds of broken links you need to check for and fix:

- Links *from* your site
- Links *to* your site

AltaVista Search is particularly handy for helping you maintain links to and from your site in two situations. First, if you discover that one of the links from your site to another one no longer works, you can use AltaVista Search to track down the new address. Just set up a query for any information that you know about this link—content, filename, or anything else.

Note *If you see in the logs for your site that people have been trying to reach nonexistent pages at your site, note the wrong address and use AltaVista Search to track down the Web page with the mistake. For instance,* link:yourdomainname/wrongaddress.html *will return a list of all Web pages that have the URL* yourdomainname/wrongaddress.html *embedded as a hypertext link.*

Second, if you have to reorganize part of your Web site or move to a new service provider, you can use AltaVista Search to find many pages on the Internet that linked to the old pages or old site. (Then, of course, you have to contact everyone who linked to you and tell them about the new address, but you knew that!)

The query *link:oldaddress* will provide you with a list of every Web page (including your own) which has links to a Web page at the URL "oldaddress."

Remember *Keep in mind that when you make changes in your pages, the results are not immediately visible at AltaVista Search. Only after Scooter has visited your site again will the changes affect the contents of the AltaVista Search index.*

Overall Site Inventory

Since people will frequently access pages in your site from somewhere on the Internet (as opposed to from your home page), you should occasionally take a look at how your pages look when they're accessed individually.

In general, you should inventory your site for the information that appears with AltaVista Search results, as follows:

- HTML title
- Abstract (the default is the first words in your document)
- Date last modified

You should apply all of this information to *every* page in your site.

You can use AltaVista Search to help you inventory your Web site in two ways. First, you can use it to obtain a complete list of the pages at your site. If the domain name of your Web site is *acme.com*, the query

```
host:acme.com
```

will return a list of the pages at your site.

If you have less than 200 pages, then Simple Search will suffice to provide a complete list. If you have more than 200, use Advanced Search, leaving the Ranking box blank, so you can get many more of them.

Such a results list can show you where you've overlooked things—for instance, using the same title for several different pages. The date on each entry can also show you which pages are old and may need to be updated.

Second, you can use AltaVista Search to find pages that shouldn't be there—versions of pages that aren't yet supposed to be public and that you thought there weren't any links to. If you find one of those, you can do another search to see where the link (or leak, in an information sense) is coming from. If the file is called *problem.html,* you'd search for

```
link:acme.com/problem.html
```

That query should return any and all pages that link to your page named *problem.html.* You can then take steps to remove that link, rename the file you don't want people to see, or just move it back to your local system and keep it off the server until it is ready to be seen.

HOW ALTAVISTA SEARCH FINDS NEWS ARTICLES

AltaVista Search finds Usenet (or network news) articles differently from the way it finds Web site URLs. To find Web site URLs, AltaVista Search sends Scooter out to roam the Internet. With Usenet articles, AltaVista Search has the articles delivered right to the door. It's like the difference between going to the grocery store or having the goods delivered at your home.

Usenet articles traverse the Internet by being forwarded from server to server across the Internet—the news server at AltaVista Search receives the full "feed" of 20,000+ newsgroups. The AltaVista Search news server maintains and indexes all current newsgroup articles, and, on request, serves the full text of these articles to users. Because new articles appear and old articles expire all the time, the news server at AltaVista Search is in fact quite busy, even though the total index of Usenet articles is much smaller than the Web index.

EXCLUDING NEWS ARTICLES FROM ALTAVISTA SEARCH

If you have any qualms about people finding a newsgroup posting of yours, you can exclude the article, just as you can exclude information in Web sites. All you have to do is enter a command—*X-No-Archive: Yes*—in the header of your message to ensure that AltaVista Search will not index it.

The following steps will prevent your newsgroup posting from being archived by AltaVista Search:

1. As you create your posting, include a line like the following at the very top of the message (skipping no lines):

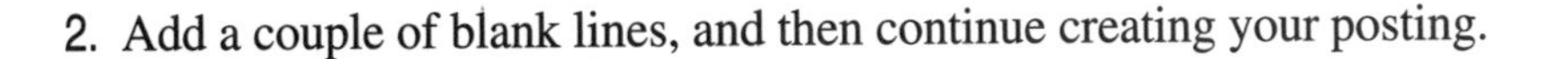

```
X-No-Archive: yes
```

2. Add a couple of blank lines, and then continue creating your posting.

AltaVista Search does not index any postings that contain the field *X-No-Archive: Yes* in the header. This field is meant to be on a line by itself in the header of the message, like the *From:* field (or the keyword or summary fields from the previous examples). However, because some users cannot include the header in the standard way, AltaVista Search also recognizes this command if it is found anywhere in the header, anywhere in the first few lines of the message body, or anywhere in the last line of the message.

Keep in mind that, even with *X-No-Archive: Yes* in your document, many people will be able to find your article on the Internet and it will probably exist, in one place or another, for years. If you post an article in Usenet, it's entirely possible that it will resurface at the least opportune time. If you have information that you really don't want to make public, don't post it to Usenet.

Using AltaVista Search for Searches of Your Site

After you have completed cleaning up your site and you feel good about how all your pages appear on an AltaVista Search match list, you might want to add a link to your home page that is a bookmark of a query for useful subsets of your pages. In other words, the anchor might read "Quick index of all pages about ophthalmology at this Web site," and the hyperlink would be the URL of an AltaVista Search search for:

```
+url:yourdomain.com +ophthalmology
```

Keep in mind that when you do this, the user "leaves" your site and connects to AltaVista Search. But the results list that automatically appears (without the user having to fill in a query box) is a set of choices, all of which are within your site.

SOME POINTERS TO OTHER USEFUL CHAPTERS

In this chapter, you've learned how to use AltaVista from the perspective of an information provider, including how to optimize pages and how to submit them effectively. So, where to next?

- Stop by Chapter 4 to learn how to use AltaVista's awesome Refine feature.
- Visit Chapter 5 to learn about AltaVista's Advanced Search.
- See Chapter 6 to learn about doing Simple and Advanced Usenet searches.
- Peruse Chapter 9 for sample searches and nifty search ideas.
- Check out Chapter 10 for the AltaVista story.

Discovering the AltaVista Search A to Z Reference

The AltaVista A to Z Reference offers a collection of sample searches, examples of real-world searches, and a number of tips to make your searches more effective and successful. Primarily, the Reference provides examples and samples to help you generate new ideas on how to search and which techniques to use in different situations. In addition, it gives you a broad idea of the range of topics available through AltaVista. While it doesn't attempt to show everything—or even a significant proportion of everything—either on the Internet or in newsgroups, it does provide a representative sample of the diversity of information you can easily find, demonstrate techniques you can use to create your own effective searches, and show how other people use AltaVista.

INFORMATION IN THE REFERENCE

The A to Z Reference provides an alphabetical collection of successful and effective AltaVista searches and a compendium of information that people have found using AltaVista. The items in the Reference are based on comments that AltaVista users have volunteered as well as some of the most popular topics and categories of information AltaVista has uncovered. As you thumb through, you'll see subjects from acquaintances to colleges to medical information to sports, and you'll learn to use AltaVista to find the information that you most want to locate.

The A to Z Reference provides something better than a complete or comprehensive listing of information on the Internet or in the newsgroups. In lieu of a reference of what's been found, consider the A to Z Reference an encyclopedia of *how* to find information. When you look under a given letter, you'll find categories of information, sample searches, explanations of searching techniques, and even personal stories from other AltaVista users about their experiences.

If you'll pardon the cliché, the A to Z Reference should teach you to fish on the Internet, rather than giving you fish. Frankly, the shelf-life of an Internet fish isn't terribly long—better to get it fresh. That is, all printed catalogs of Internet resources tend to become dated quickly. More information is added to the Internet every day, and a large amount of the existing information is updated regularly. In fact,

AltaVista's Web crawler (Scooter) collects 10 million new sites per day, and about 20,000 new sites are submitted manually each day. This environment of constant change makes the A to Z Reference even more useful because it provides examples of strategies on how to search and how to hone in on the information you really need.

A TO Z REFERENCE ORGANIZATION

The A to Z Reference provides an alphabetical listing of select categories and searches. Within the Reference, information is broken out by letters, just as the index of this book is divided into letters. Under each letter, you'll find several topics, chosen because they were inherently interesting or because they provided a good example of unexpected ways to search, or because someone told the AltaVista developers about their positive experience searching for that topic.

Each topic contains information about relatively obvious applications, such as searching for recipes containing *apples* (Simple Search **+recipe +apple**). Additionally, we offer further suggestions, such as finding recipes to use up the apples you have on hand, even though you don't have any cinnamon or sugar (Simple Search **+recipe +apple -cinnamon -sugar,** which yields about 4000 hits and an unknown number of servings). In some of the search examples, you'll find cross-references to other search topics within the A to Z reference. These cross-references not only will give you additional search ideas, but also will help show you just how "webbed" Internet information really is.

Also, throughout the Reference, you'll find boxes containing actual user stories. These anecdotes show how other AltaVista users like yourself have been able to apply AltaVista searches to their own informational needs. As you read through these, you're likely to see several searching techniques that you could use for your particular informational needs, and even strategies for uncovering information in ways, and on topics, you hadn't previously thought of.

WHY SHOULD I USE THE A TO Z REFERENCE?

If you've already read chapters 2 through 8, you've undoubtedly learned how to successfully search using AltaVista. You can now search for appropriate terms, review the list of results, browse through the titles, and find your way to the information you need. However, as Mark Twain noted, "The difference between the right word and almost the right word is the difference between lightning and lightning bug." The A to Z Reference provides examples, anecdotes, and guidance to help you immediately identify your personal lightning bug.

Seeing successful searches, for example, on marketing plans or women's rights may help you improve your own searches for entirely different information. Possibly you hadn't even dreamed of searching for some particular snippet of data, or you'd tried to find something but overlooked one key issue. The samples and examples throughout this section will help you ease into improving your own searching techniques and build on the fundamentals you mastered in the previous chapters of this book. AltaVista is a powerful tool that you can use in unique ways to unearth information about almost anything. We've structured the A to Z Reference both to show you how others have used it and to give you ideas on how to apply its capabilities to your own needs.

Just one more note: Some of the more obscure information on the Internet might well go unnoticed if people didn't have access to AltaVista. Let your imagination run wild when trying the program. It works.

HOW SHOULD I USE THE A TO Z REFERENCE?

Start by browsing through the Reference just to see the topics and read the user stories. If something catches your eye, read it. That'll be the easiest and most enjoyable way to dive in. We didn't design this reference to be read from cover to cover, although you'd learn the most that way. After you're familiar with the content and have flagged a few interesting items, go back and reread those, possibly while sitting in front of the computer and trying some of the examples out with your own search topics.

Don't feel any obligation to try every example or to work sequentially through the text. Many of the examples are likely to be appealing while others will make you wonder where some people find so much free time. Think of the examples as sample how-tos, not sample what-tos. And remember, AltaVista Search is an index of the content of the Web and Usenet newsgroups—if there is a topic you want to know more about, AltaVista Search finds where the information resides and brings it back to you.

A TO Z REFERENCE CONVENTIONS

The A to Z Reference consistently uses some simple typographical conventions to make it easier for you to spot the information you want. Throughout this section, the actual search query you'd type is shown

```
on a line by itself, like this.
```

The reference information itself will be the main flow of text, broken up with alphabetical dividers and headings indicating each category. All of the user stories and anecdotes will be in boxes, located near the topic they're related to. Some of the user stories have been edited for length or paraphrased, but all reflect the essence of authentic comments about AltaVista.

A

acquaintances

Suppose you're trying to track down an old friend or someone who owes you money. Or perhaps you're looking for a long-lost buddy. A good place to start is with AltaVista's People Search. Just click the People Search tab and type in all the information you know in the Switchboard People Search screen—last name at a minimum. If that doesn't do it, head back to AltaVista with an Advanced Search for the first and last names. For example, if you're looking for Albert Einstein, search for:

```
Albert NEAR Einstein
```

on both Usenet and the Web. If that doesn't bring up anything useful (perhaps your friend doesn't have a Web site and doesn't post to Usenet), search out:

```
"white page*"
```


and try some of the other dedicated White Pages services. These services are similar to your local phone book, but are not comprehensive or necessarily up-to-date.

In Search of African Philosophy

In a comment about AltaVista Search, Jesse wrote:

> *I am always trying to get faculty [at my university] to use the Internet. Recently a colleague came in and told me he was interested in finding some information for his philosophy course. However, he said he doubted there would be anything on African philosophy. He was surprised to find hundreds of hits and ran off to get a disk to download several articles.*

—Jesse Silverglate

annual reports

Need to check up on the status of a company? Need to figure out if you should unload that stock you inherited? Pick a company and search away for +*"annual report"* and the *company name*, for example:

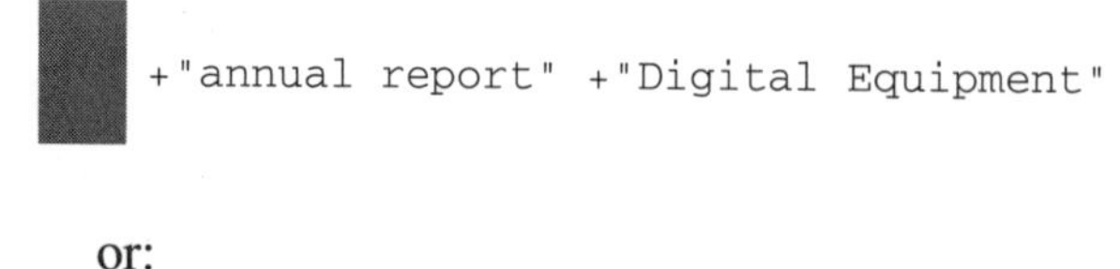

```
+"annual report" +"Digital Equipment"
```

or:

```
"earnings report" +the company
```

These searches give you information about the specific company as well as the reports they publish on the Web. If you're looking for more current information, try a Usenet search for:

```
+newsgroups:invest
```

or, to be a little more specific, search for *+newsgroups:invest +companyname*, as in:

```
+newsgroups:invest +IBM
```

anonymity

The generally impersonal nature of computer networks coupled with the ease with which people can anonymously post messages has led to all kinds of interesting discussions that you can find through a Usenet or Web search on:

```
anonymity
```

or:

```
anonym*
```

This search provides you with all sorts of information and discussions about the pros and cons of anonymity on the Internet—interesting, not to mention a very hot topic.

If you specifically want information about anonymous email—for example, to participate in self-help groups without fear of being identified—try a search for:

```
anonym* email
```

In Search of Answers

In reference to a frequently asked question on a newsgroup, Dave wrote:

> *No. I'm not gonna go through that again. This is the dozenth time I've written about this, and AltaVista came up with 50 or so posts with "mailto" and "subject" in the last three months. Did you look there?*
>
> —Dave Salovesh

art (images)

If you're interested in finding information on art (like images, pictures, or paintings), AltaVista Search is the place to look. Because of the vast amount of information on this subject available on the Internet, entering common terms like *images, art,* or *paintings* won't be too useful—these searches result in thousands of hits that would take forever to wade through.

Instead, you should narrow your search, for example, by the specific artist, like this:

```
pictures Escher
```

or, depending on your taste:

```
pictures Dali
```

or even:

```
pictures "Van Gogh"
```

If you're looking for information on a type of art or a specific art period, you could try:

```
pictures impressioni*
```

or:

```
+pictures +renaissance
```

In Search of Astronomy

In a comment about AltaVista Search, Deane wrote:

> *My father is interested in astrophysics and astronomy, and wanted some information on a particular star called Betelgeuse. When I searched JPL's and NASA's home page, I couldn't find anything. Next step was to search using Yahoo and WebCrawler, and they didn't find anything.*
>
> *Then I decided to use AltaVista, and on the first try you found it. Now if I have to find something on the Internet, I go straight to AltaVista for it.*

—Deane D. Davis

B

baseball cards

"Take me out to the ball game…." Well, AltaVista Search can't take you to the ball game, but it can help you find that rare baseball card you've been looking for.

Suppose you want to add the Ty Cobb baseball card to your collection. You could search newsgroups by entering:

```
+"Ty Cobb" +card* +"for sale"
```

This search gives you several listings of cards and places to buy and sell originals as well as reprints. You can get more specific, for instance, trying to track down the extremely rare card from his rookie year by entering:

```
+"Ty Cobb" +card* +"for sale" +rookie
```

With this search you'll find information on the card, its condition, and the price.

basketball

If you're a basketball fan, AltaVista can find loads of information about it for you. You can do a general search on the topic, like:

```
basketball
```

Doing this Simple Search provides you with a wide variety of information, ranging from team status to scores to most valuable players. If, however, you've entered a bar bet with your basketball buddies, you might want to hone your search a bit and find out the latest scoop on the team standings. For instance, you could enter the following Simple Search:

```
basketball handicapping
```

Also, searching for teams by name is generally productive, as in:

```
"Chicago Bulls"
```

Or you could search for officially sanctioned sites by entering:

```
basketball official
```

This search gives you not only the officially sanctioned sites, but also information on official basketball rules. If you don't want the basketball rules included, you could enter:

```
basketball official -rules
```

Or, if you *only* want information on official basketball rules, you could enter:

```
basketball official +rules
```

Finally, if you're into collecting basketball memorabilia, you could do a Usenet search to link up with other collectors, as in:

```
newsgroups:rec.collecting.sport.basketball
```

beekeeping

If buying your honey some honey isn't enough, you could always grow your own—more fun than a barrel of…er…bees, we guess. Just a Simple Search of the Web for:

```
beekeeping
```

should get you started. On the other hand, you might try:

```
bee sting
```

to find everything from treatment for bee stings to using bee stings therapeutically.

bicycling

A search for:

```
bicycling
```

not only provides links to information about alternative transportation and columns about races, but also a number of other informational links. You can even find out just how people decided to perch on top of these two-wheeled contraptions by searching for:

```
bicycle history
```

for a full history, including some nifty pictures.

biochemistry

You can use AltaVista to find out the very latest information about a rapidly evolving topic. For example, suppose you are researching the details of the protein folding process. You'd like to determine if it is predictable and, hopefully, controllable. You have a hunch that the mathematics of chaos theory could be a useful tool here. You could first search the Web for:


```
"protein folding" chaos
```

Or you could refine the search even more by doing an Advanced Search to limit the topic by date. Remember, if you're looking for the latest and greatest information on a topic, you could try using the same query to search Usenet newsgroups, which often provide you with the most recent information available on a topic.

black history and literature

Suppose you want to put together an anthology of black history and literature in electronic form to distribute to your class. You are specifically interested in works by Frederick Douglass and W. E. B. Du Bois. You could do an Advanced Search for:

```
("Frederick Douglass*" OR "du bois") AND (etext OR "electronic book")
```

You could focus the search even more, for example, by limiting the search to organizations that make electronic texts available over the Internet. If you're familiar with Project Gutenberg, you might add the name of the organization to the ranking box, like this:

```
Gutenberg
```

Now Project Gutenberg's index appears at the top of the list, and many of the other items are pointers straight to specific works by Douglass and Du Bois that are available free online from Gutenberg.

blind

Suppose you're looking for information on text-to-voice converters, which make it possible for the blind to navigate the Internet and "read" electronic texts. Graphical user interfaces—Windows in particular—threaten to lock them out. Hence many people are very concerned about and involved in efforts to make Windows applications accessible to the blind. For updates on what is happening in that area, search both the Web and newsgroups for:

```
+blind +Windows
```

Increasingly, you'll find more information about this area in general with a search for:

```
"vision impaired"
```

Or suppose you have a blind friend who enjoys knitting and you would like to buy her a gift. You could do a search of the Web using:

```
+knitting +braille
```

and find a variety of resources for blind knitters, including a book in braille from the National Braille Press.

brewing

Thinking about brewing your own drinks? Try reading up on different opinions and ideas. Do a general Usenet search for:

```
newsgroups:brewing
```

or:

```
newsgroups:tea
```

If brewing your own beer or tea just hasn't been quite as successful as you'd hoped, you might look for:

```
newsgroups:brewing recipe
```

or:

```
newsgroups:brewing troubleshooting
```

and find specific brewing ideas. Alternatively, you could head over to the Web side of things and do a quick search on:

```
zymurgy
```

business

You can't just search for *business,* because that word is so common it is ignored. Phrases that include the word *business*, however, are still a good way to go. For example, if you're trying to learn how to write a business plan, try:

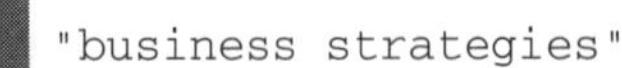

```
"business strategies"
```

or:

```
"business plan" +"how to"
```

Also, the U.S. Small Business Administration (SBA) has all kinds of good information that you'll find by adding *+host:sba.gov* to your search, as in:

```
"business plan" +host:sba.gov
```

Or suppose you're in the market for a new computer and want to check the Better Business Bureau records on the business from which you want to make a purchase. All you'd do is enter:

```
"better business bureau"
```

Using this search string, you can find lots of Better Business Bureau sites that can help answer your questions. Or better yet, add the name of the region or community, like this:

```
"better business bureau" +Tulsa
```

You can even find out what other people have said about the company or institution of your choice by searching for *company name +host:angry.org,* as in:

```
ABC Computers and Parts +host:angry.org
```

Finally, if you know the geographic area of the business and what it does, but just need some contact information, don't forget to try AltaVista Business Search, addressed in Chapter 3.

C

camping

So, you're ready to leave the desk behind and head out to the wilderness? You'll need some camping gear to do so, right? Buy it online with a search for:

```
camping gear
```

With this search, you'll find links to sites where you can purchase equipment as well as links to comments on different gear. Of course, you'll probably need some advice and instruction first, so try:

```
backpacking camping "getting started"
```

This will give you a slew of links with useful advice—however, as ever, consider the sources carefully.

canines

There's all sorts of information to be found by doing a Web search for:

```
canine
```

You'll find information from canine breeds to training to housebreaking. Or you could even find information on specific canine topics, like:

```
canine genetic diseases
```

which you might want to check out before you buy that cute puppy in the window.

As always, though, if you're looking for information about a specific dog breed, enter the specific breed's name—no, not Sparky or Rex, but:

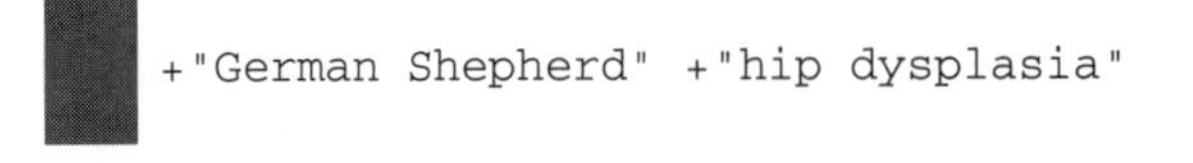

```
+"German Shepherd" +"hip dysplasia"
```

caning furniture

That old chair's not holding up very well? Need to replace the cane on the seat? A Simple Search for:

```
caning furniture
```

gets you there. You'll find all sorts of businesses that can help you with your caning needs. Or you could limit the search to just your area, using +*yourcity*:

```
caning furniture +Houston
```

cars

Your old heap just left you stranded again? Spring for a new one! You can use AltaVista Search to find current information about buying cars—the latest reports, repair history, buying guidelines, and on and on.

Besides the obvious Simple Usenet searches for:

```
cars automobiles
```

specifically targeted searches can be really useful. If you're seeking buying advice, turn to the AltaVista Advanced Search and put in the search query:

```
(car OR automobile) AND buying
```

and in the Ranking field:

```
used
```

Of course, if you want information about new cars, substitute *new* as the ranking criteria. Likewise, substitute the make or model if you'd like to pull up information about a particular car. Switching over to the Web searches within AltaVista Search and putting *buying guide* into the Ranking field can also help. Note that you can substitute *repair** as the ranking criteria if you decide it'd be easier to fix your old car than to buy a new one.

If you still need to sell the old heap, you could place ads in local papers and post to classified ad sites on the Web and to newsgroups. But if at all possible, you'd like to make a very quick sale; you'd like to find someone right away who is looking for the very thing you have to sell. You could try an Advanced Search of newsgroups, as here:

```
(wanted OR need) AND ((VW OR Volkswagen) AND Jetta)
```

Or you could try the same search on the Web. Toward the top of the results list is a home page that's nearly perfect.

CD-ROM development

Suppose you have an idea for an interactive CD-ROM designed for kids. A friend suggests that you get in touch with an animator, Derek Lamb. You know nothing about the CD-ROM business, and don't know who this "Derek Lamb" character is. You can begin with a Simple Search of the Web:

```
+animator* "CD-ROM" +"Derek Lamb"
```

Or you can broaden this search a bit by searching for just the animator's name, like this:

```
"Derek Lamb"
```

Tada! Among other information, this search results in a directory of animators, where you can get the street address and phone number of Lamb's company.

chemistry

Searching for chemistry information at any level? No problem; just tune into Usenet and try:

```
chemistry
```

but be sure to specify additional information, like *reactions, physical,* or *journal.* Otherwise, the random information that uses the word *chemistry* might overwhelm you.

Likewise, the Web contains a wealth of chemistry information for all needs. Try:

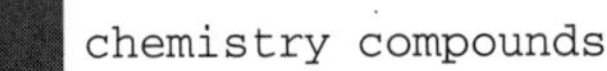

```
chemistry compounds
```

or:

```
nuclear reactions
```

Or if you're experimenting with human chemistry, try:

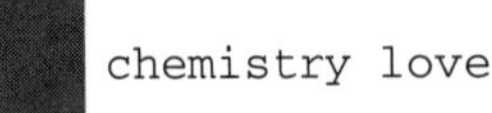

```
chemistry love
```

chess

If you are looking for a chess teacher for yourself or your child, just use Simple Search and enter all the related words:

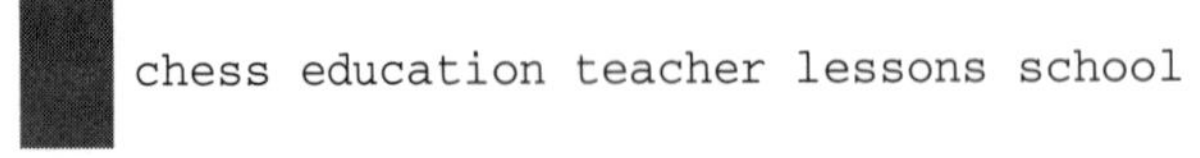

```
chess education teacher lessons school
```

You may get many thousand responses, some of which look useful, but there are far too many to wade through. To focus more closely, try the likely phrases:

```
"chess teacher" "chess school"
```

With this search string, you'll get far fewer but more useful results. Or you could even leave the same query in the box and switch to Usenet to find related newsgroup articles, as follows:

```
newsgroups:chess teacher school
```

You'll get thousands of responses, but the top ten or so include useful discussions about chess education and particular teachers and schools. Just what you're looking for!

You can even use AltaVista Search to find out the name of a chess opening you have never seen before—simply type the moves and search on both the Web and in the newsgroups, and chances are great that if that sequence has been played often before, some of the postings include its "official" name. For instance, enter:

```
e4 c5 Nf3 d6 d4 c5xd4 c5:d4 Nxd4 N:d4
```

Because of variations in notation, you need to enter both "x" and ":" for capture moves.

Keep in mind that while this only works for games that use the English abbreviations for pieces in their notation (such as "N" for knight), that represents the vast majority of games on the Internet. Also, note that since you are not specifying the order in which the moves are played, you would also find games in which the move order is transposed. In any case, this technique is quite valuable for studying an opening, researching how to annotate a game, or finding games similar to one that you played so you can look at alternative lines.

childcare

If you—like many other people—need to find someone to care for your child or children while you're at work or meetings or school, use AltaVista to collect advice and checklists. Try a search for:

```
choosing good safe childcare
```

Of course, you could also try alternatives, like:

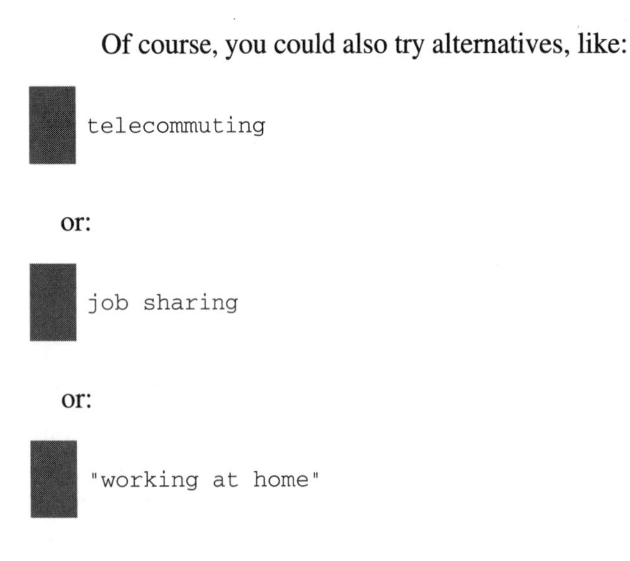

```
telecommuting
```

or:

```
job sharing
```

or:

```
"working at home"
```

child development

Yes, AltaVista Search can even help answer your questions about child development. Wondering what solid food is best to start your baby on? Want to know why your baby drools so much—or more likely, when it's going to stop? Curious about when your child's babbles will turn into real words? If you're just looking for general information about child development, try:

```
"child development" process
```

Using this search string, you'll find lots of information on this subject; however, you'll probably have better luck with something more specific, such as:

```
"child development" premature
```

or by focusing on specific issues, such as:

```
language development children
```

clip art

If you need some clip art to spruce up your Web site or that newsletter you got drafted into doing (you didn't *volunteer* to do it, did you?), try:

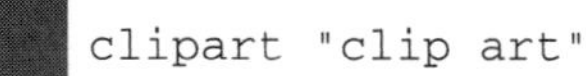

```
clipart "clip art"
```

If you're interested in only the *free* clip art, try:

```
clipart "clip art" free
```

Both of these examples use clip art with and without a space, because different people write it in different ways on the Internet.

Remember, just because you find it on the Internet doesn't mean it's yours to use. Be sure when you're pilfering a nifty clip that it's really out there for you to take. See the "copyright issues" section of this Reference for search ideas on that topic.

In Search of Coconut Oil

In response to a question about finding sources to buy coconut oil, Chuck wrote:

> *My normal advice is: Try using AltaVista's Advanced Search by first going to*
>
> *http://altavista.digital.com*
>
> *Then click on Advanced Search, read the help, bookmark the page. You can then use AND, OR, NEAR, NOT, and nested parentheses and quoted strings with some wildcards.*
>
> *The results of a search for "coconut oil" & order & popcorn:*
>
> *Word count: coconut oil: about 2000*
>
> —Chuck Demas

9

colleges (and universities)

So, you say you want to go (back) to college? And you say you've visited every bookstore in town only to find those huge books that list a bazillion colleges, supposedly to help you find the college of your dreams? Yeah, right. AltaVista Search can help!

If you already know the name of the college you want to attend, you can just enter the name, like this:

```
"Washington University"
```

or:

```
"Oklahoma State University"
```

If you only know the general area in which you want to live and need information on the colleges in that region, you could enter a search like this:

```
+Seattle colleges universities
```

This search gives you colleges and universities in the—you guessed it—Seattle, Washington, area. If you enter a search like:

```
+Portland colleges universities
```

you'll get search results from Portland, Maine, and Portland, Oregon. You could refine even further by entering, for example:

```
+Portland Oregon colleges universities
```

comic books

If baseball cards aren't your bag, you could search for the collectible of your choice. If it's comic books, you can search the Web to find pages put together by fans of the same series as you, or you can search the newsgroups for particular back issues to fill out your collection. For example, search the Web for:

```
+marvel +2099
```

and you find hundreds of Web pages dealing with Marvel Comics' 2099 series. (There is no need to include the word *comic* in your search, since *marvel* is enough to specify what you want.)

You can also search newsgroups, using:

```
+marvel +2099 +"for sale"
```

and find hundreds of recent postings listing issues in that series that are for sale. To narrow the search even more, try Advanced Search and enter the numbers of the exact issue(s) you want to buy, like this:

```
marvel AND (X-nation OR Xnation) AND (2 OR 10 OR 15)
```

If you don't find what you want this time, bookmark the search and try again later. Sometimes the information or issue you want isn't yet available on the Internet.

In Search of Computers

In a comment about AltaVista Search, Mike wrote:

> *I really appreciate DEC creating AltaVista. I feel I could give you one "success" story a week about how AltaVista has helped me do my job more efficiently. I find I use AltaVista at least once a day and I'm frequently amazed at the results. Case in point—we have an old HP plotter that we are trying to connect to a Macintosh. We were having trouble finding the pinouts for the cable until I logged in to AltaVista and typed:*
>
> *plotter Macintosh cable*
>
> *After days of frustration, AltaVista gave me the answer in seconds—outstanding!*
>
> —Mike Honeycutt

9

communicating with old school friends

The first choice on opening up communication with your old school buddies is to use AltaVista's People Search (covered in Chapter 3) to search them down by name and location. If you don't know the location and your friends have relatively common names, however, you might need to head back to the AltaVista Simple Search page and search for the name and all of the other identifying characteristics you can think of. For example, you might search for your school name and the friend's name. Additionally, just search for the school and check for alumni lists.

contests

Want to enter some contests? No problem! Either on the Web or on Usenet, search for:

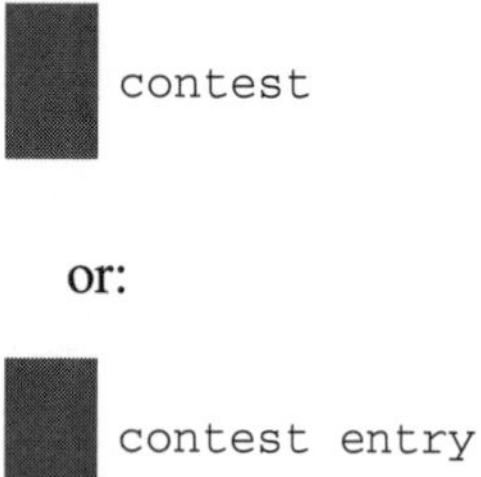

```
contest
```

or:

```
contest entry
```

Doing an Advanced Search and restricting the range of dates to the last few days might help identify the contests before everyone else has already entered and won.

cooking

And you thought regular cookbooks had loads of recipes—wait until you use AltaVista Search to find them! For the constantly changing collection of Usenet recipes, search for:

```
newsgroups:recipes
```

If you want something specific, try, for example:

```
newsgroups:recipes bisque
```

or:

```
newsgroups:recipes chicken
```

If you're interested in getting a broader perspective than just recipes, try:

```
newsgroups:baking
```

or:

```
newsgroups:cooking
```

and throw in the items that you want to cook, such as:

```
newsgroups:cooking oysters
```

Switching over to the Web side of things opens up another world of recipes, but you'll want to have an idea of what you're cooking so your search results list isn't too overwhelming. Entering:

```
recipe +chicken tomato* pepper
```

or some such is a good way to clean out the refrigerator. If you just can't live without that airline food, try:

```
cookbook "American Airlines"
```

If you're a vegetarian or have specific food preferences, you could create a search and bookmark it for later, as in an Advanced Web Search for:

```
NOT (mushrooms OR oysters OR liver or "brussels sprouts") AND
```

then just put the food you do want to eat at the end of the query before you run the search.

copyright issues

AltaVista Search can help you do some copyright research. For example, you can find out about copyright laws, what constitutes infringement, and what to do if someone infringes on your copyright. You could try a broad search, like:

```
copyright
```

This search query results in bunches of hits—probably too many to be manageable. To limit your search on this topic, you could add words to *copyright* in the query field, like this:

```
copyright law
```

This search gives you specifics on how many of the sites on the Internet infringe on other people's rights. Or for more information about infringement, try:

```
+copyright infringement
```

For a broader perspective, you might try:

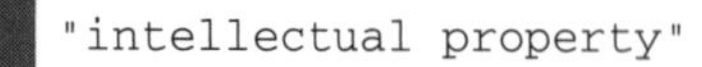

```
"intellectual property"
```

crafts

So, someone you know and love is into crafts? Send them to AltaVista to learn where to start turning those toilet paper holders into real money. For example, try:

```
sell* craft* fair* project
```

Here, you'll find lots of tips and ideas, but be sure that you can recognize:

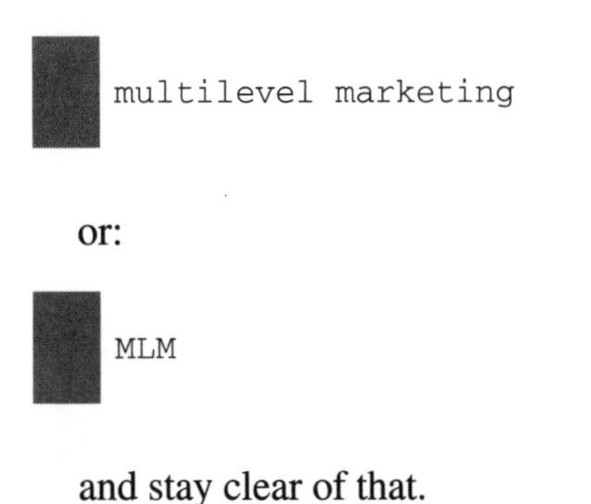

```
multilevel marketing
```

or:

```
MLM
```

and stay clear of that.

D

dance

Looking for a weekend pastime or an after-school activity for your kids? Try dance! You could start with a broad search, like:

```
dance
```

to see a wide variety of information on dance. A quick search for these topics:

```
country line dancing
macarena
ballet
dance +instruction
```

would get your toes tapping with information about the dance forms as well as information on where you can take lessons. You can also find information on more rare types of dance, such as:

```
dance celtic
```

if you're researching historical dances, or even:

```
dance chicken
```

if you're looking for information about the famous "chicken dance" you've seen at local Oktoberfests.

> **In Search of Dance School**
>
> In response to someone looking for the London School of Contemporary Dance, Victor wrote:
>
> *You didn't search very hard. This took me literally less than a minute with AltaVista (http://altavista.digital.com/).*
>
> *Search for **London School Contemporary Dance.** It's at:*
>
> *http://www.ecna.org/placeds/lcds.html*
>
> —Victor Eijkhout

daycare

Yes, AltaVista Search can even help you find a spot for your tot! From choosing to using, just search the Web for:

```
daycare
```

and you'll find useful information on how to select a daycare provider and how to help your little one adjust. Also, you could add your location if you're looking for a specific provider in your area, as in:

```
daycare +Dallas
```

diet/lowfat references

In addition to the neat cooking ideas and recipes you can find with AltaVista Search (see the "cooking" section of the Reference), you can also find good diet and lowfat references. For example, you can do a Usenet search for:

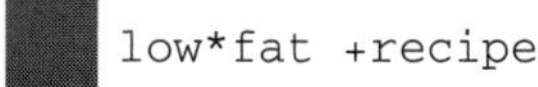

```
low*fat +recipe
```

This search yields some pretty handy lowfat recipes, while:

```
cholesterol
```

leads to broader discussions, including information about cholesterol in food as well as its effects on humans. Over on the Web, you can find more information about a lowfat diet, eating well, and exercise. For starters, try:

```
"good health"
```

donations

Many people want to give to a good cause, but it's often hard to tell the difference between a good cause and a scam. AltaVista Search can't decide that for you, but it can point you to some causes that can put your money to good cause. Try a search for:

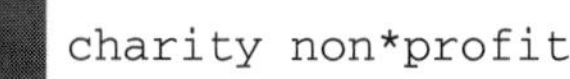

```
charity non*profit
```

or just the cause that you're interested in, such as:

```
"American Heart Association"
```

By the way, the * in non*profit ensures that you'll find both the non(hyphen)profit and nonprofit organizations, depending on how they're written.

drums

Tappity, tappity, tap. You can find all sorts of information about drums on the Web, just by entering:

```
drum*
```

This general search gives you links for drum lines, drums (to play in a band), drums (to beat in the wilderness), and drummer figurines, among other things. If you're looking for more specific information, try entering, for example, specific types of drums, like this:

```
timpani
```

or:

```
marching snare
```

Entering these same searches in Usenet can also provide interesting information about drums. You can find all the opinions and advice you want, even information on how to tune marching snares ("tight" is the word on the street).

Also, you can search for drum-related topics, like:

```
"Green Drummer"
```

which leads you to a newsletter from an environmental organization. Also:

```
rhythm -"rhythm method"
```

might be useful if you march to your own drummer.

E

education

AltaVista Search allows you to search for a huge range of education-related topics. For example, suppose you believe that your second grader is gifted and you want to find a private school that will give her the opportunity to develop to her fullest. You need a school that is within commuting distance of Boston. All you have to do is search the Web with:

```
+"gifted program*" +"private school*" +Boston
```

Or suppose, while channel surfing on TV, you accidentally came upon the end of an educational program on comets. You notice that there's supplemental information on the Web but didn't catch the URL. You can still find the information on the Web by entering a Simple Search, such as:

```
comet* PBS
```

You'll not only be able to reference the supplemental information that the TV program mentioned, but you'll also find the broadcast schedule for the other programs in the series, an online teacher's guide, opportunities for students to interact with scientists, and dramatic space photos that you can download.

emergency planning

Just in case you were wondering whether your company or community is prepared for emergencies, you can check out other places' emergency preparedness plans, guidelines for planning, and similar information with a Web search for:

```
emergency planning
```

Or, if you're interested in helping a nearby town struck by disaster, you can search for:

```
disaster recovery
```

and find out ways you can help.

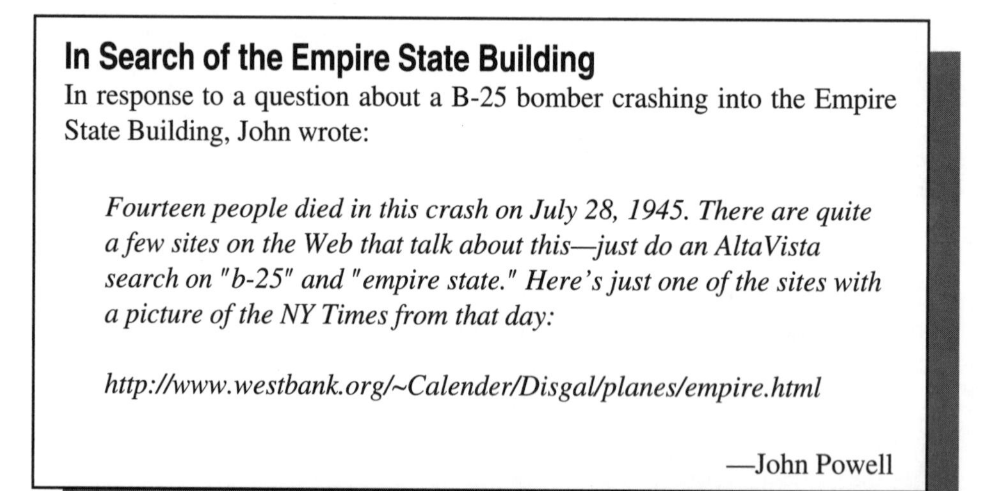

In Search of the Empire State Building

In response to a question about a B-25 bomber crashing into the Empire State Building, John wrote:

> *Fourteen people died in this crash on July 28, 1945. There are quite a few sites on the Web that talk about this—just do an AltaVista search on "b-25" and "empire state." Here's just one of the sites with a picture of the NY Times from that day:*
>
> *http://www.westbank.org/~Calender/Disgal/planes/empire.html*

—John Powell

environment

Suppose you are concerned about the environment and in particular about the pollutants discharged from industry in your community. For starters, you'd like to know what the U.S. Environmental Protection Agency and others have to say about safe levels of contaminants—say, copper—in drinking water. All you have to do is enter an Advanced Search, such as:

```
(EPA OR Environmental Protection Agency) AND ((wastewater OR
"waste water") OR discharge) AND copper
```

ergonomics

Spending so much time surfing the Net that your hands hurt? Try:

```
ergonomics "carpal tunnel" RSI
```

This search gives you links to information on carpal tunnel syndrome (also called Repetitive Stress Injury, or RSI) as well as information on how to diagnose it, find help for it, and continue to work with it. After you've read up on it, try:

```
ergonomic office furniture
```

for information on office furniture that can help prevent RSI disorders and help improve your work environment.

The same search over on Usenet is also worth trying, possibly with the addition of:

```
newsgroups:human-factors
```

which delves into a number of ergonomics-related issues.

espresso (or cappuccino)

You need to keep going and just gotta have a pick-me-up? The daily grind got you down? Try a search for:

```
espresso cappuccino
```

This Simple Search gives you information about espresso and cappuccino, the different kinds, and the trends. If you want to do more than just read about it, try adding your location to the end, as in:

```
espresso cappuccino +Dayton
```

(This works better if you live in Seattle than in Dayton, but AltaVista Search serves you the information that's out there, nonetheless.)

Switch over to the Usenet side of AltaVista Search and try a search on:

```
espresso
```

for some discussion. Or:

```
newsgroups:coffee
```

broadens the scoop...er...scope.

In Search of Employee Relations

In a comment about AltaVista Search, Ethan wrote:

> *This afternoon, I lost [my] Web tool AltaVista [through technical difficulties]. There was just no way I could access my favorite search engine/database. But I didn't realize until three hours later just how valuable AV is. I am in the process of writing my fourth book, and desperately need information on "minor details" and some citations on employee relations. I finally had to try other search engines, and not once did I find the references I was after. Only AV can find the obscure... To make a long story short, I have never "missed" something as much as I missed AltaVista. I was at a total loss. Thank heavens it's back. It was a definite Eureka! experience!*

—Ethan Winning

exercise

If you've discovered that exercise is more effective if you actually *do* it (rather than watching others do it on TV), then you might benefit from the information you can find using AltaVista Search. For starters, try:

```
exercise health fitness
```

Good for you for getting out there and exercising, but—ouch!—you say you've overdone it? Try:

```
sports medicine
```

to find information on common sports injuries, prevention of injuries, and remedies. If you find that you need to call in the big guns, search for:

```
MEDLINE
```

(yes, use all caps), which gives you the MEDLINE medical database. Or check out the "medical" sections of this Reference for more information.

F

fast food

If you're interested in finding out just how bad for you that quick snack on the road was, try a search for the nutritional analysis of the food, as in:

```
+"nutrition* information" +McDonald*
```

With this search, you'll find all sorts of information on calories and fat, and even some information on healthy choices.

fish

If you've decided that the only pet you'd ever want is quiet, clean, and doesn't require walks or litter boxes, try a fish. If you're just getting into it, start on the cheap with a Usenet search for:

```
aquarium "for sale"
```

Again, add the location if you like, as in:

```
aquarium "for sale" "des moines"
```

You'll probably need some information about how to go about it (there's more to it than just throwing fish in the sink), so try a Web search for:

```
aquarium "getting started"
```

This search provides you with an assortment of information on selecting equipment, feeding fish, and buying fish, just to name a few. (Of course, if fish as pets aren't your cup of tea, check out "cooking" in this Reference for other ideas.)

flying

If you're interested in the history of flight, you should try a search for:

```
flight history
```

foreign language and culture

Need to keep up your foreign language ability? Just focus your searches on the language or country of your choice. For example, to see only Web sites written in German, choose German from the drop-down menu at the top of the Search form. Or, if you want, you can get only sites from Germany by adding *+domain:de* to the search of your choice. For example:

```
yourtopic +domain:de
```

With this search, you could find lots of facts about Germany, including tidbits on popular tourist spots and historical sites, and, of course, information about learning the language. Because AltaVista Search covers the whole Internet, just by typing in the search string in the foreign language, you'll generally get the information and language you want.

Remember, AltaVista also provides translations for foreign language Web pages—check out Chapter 2 for a reminder.

forest fires

If you want to learn more about forest fires, their causes, and how forest management practices affect them, try a Web or Usenet search for:

```
forest management "forest fire"
```

Of course, if you're going camping or hiking anytime soon, you might brush up on:

```
preventing "forest fires"
```

Or, a Web search for:

```
forest fire simulation
```


lets you interactively experiment with some of the factors in forest fires without ever lighting a match. Exciting!

Also, take a look at the search ideas in the following section.

forestry

Suppose you're doing a research project on forestry—or even considering it as a major in college. AltaVista Search can help! For starters, search the Web for:

```
forestry
```

This Simple Search produces numerous hits, including information from university programs to environmental issues to fire prevention. Or, if you want to

find lively discussions on this topic, do a Usenet search on *forestry.* This search quickly brings up spirited debates on environmental and political aspects of forestry.

Or check out other search ideas under "forest fires" or "colleges" in this Reference.

frisbee

If your poor frisbee is in retirement (or your dog has used it as a chew toy), get some fresh ideas by searching the Web, like this:

```
frisbee "flying disk*" +games
```

This search gives you several choices for disk games, although no apparent ways to discourage the dog. Or if you're young, energetic, and fearless, try:

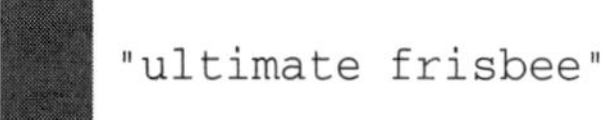

```
"ultimate frisbee"
```

which is kind of a combination of rugby, American football, and soccer. Actually, just watching ultimate frisbee can be fun, so check out a game even if you're not likely to risk broken bones.

G

gambling

If you're off to Vegas or just off to the local bingo hut, you might first try searching AltaVista for:

```
gambling
```

Doing this Simple Search provides you with information on gambling regulations, locations, resorts, lotteries, and the like. For an optimistic yet ironic look at gambling, try:

```
gambling winnings
```

This search string provides you with predictable strategies to increase your winnings (no guarantees, though) as well as loads of tax forms and tax links. If you're not likely to come up with anything for the tax man, try:

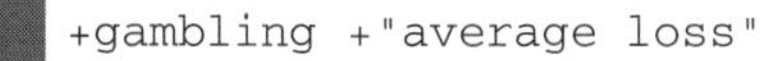

```
+gambling +"average loss"
```

Or, if you've just returned from Vegas or the local bingo hut, you might try a search for:

```
"credit counseling"
```

games

The Web is full of information about games, which you can find with a search such as:

```
board games
```

or:

```
interactive games
```

These searches give you gobs of information on games for both you and your kids. You can find out what games are available, where to get them, and how to play them well. If you're looking for specific hints or tricks, the name of the game is probably the best search string. For example, entering:

```
Scrabble
```

gives you information on everything from computerized versions of the Scrabble® game to playing tips to programs that can help you make those last three letters fit. By the way, many (if not most) computer games give you ways of jumping ahead to the next level or of beating the system (like a virtual card up your sleeve). Try:

```
+cheat +code
```

for general information, or:

```
+cheat +code +thegameofyourchoice
```

to see if someone's posted the trick. You can, of course, also try these searches in Usenet.

See the "chess" section of this Reference for more game-related search ideas.

In Search of Games

In a comment about AltaVista Search, David wrote:

I was playing around with my Rubik's cube and had forgotten an important pattern. Thinking someone may have a solution on the Web, I looked around with Yahoo to no avail. I then tried AltaVista, and there was a solutions page as the second item listed. I have found many weird things with AltaVista that I couldn't find with any other tool. Things like an old friend, a song title from the lyrics, and a satellite tracking program, all were found easily with your index.

—David Skirmont

In Search of Games

In reference to a question about codes for a Star Wars computer game, Kevin wrote:

A good way to look up codes is to use AltaVista. It lets you search through the newsgroups and most of the time the code has been posted once.

I found the code you were looking for by searching on "32x star wars code."

—Kevin McGill

gardening

The available information on the Web about gardening is growing like weeds! You can search for general information by entering:

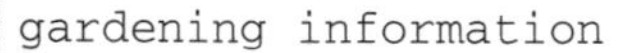

```
gardening information
```

Doing this Simple Search, you'll find all sorts of information about gardening, ranging from tomato tips to how to ensure that your gardening is environmentally friendly. Depending on your specific needs, you can narrow your search results in any number of ways. For example, if you want to find out how to control those pesky bugs that have eaten your tomato crop, you could try:

```
gardening +pesticides
```

With this search, you'll find lots of information on pesticides, organic pesticides, and even hydroponics. Or, if you don't seem to have that green thumb, you could just do a search for gardening tips, like this:

```
gardening +tips techniques
```

With this search, you'll find gardening tips for all sorts of plants from around the world. You can find out how to keep Peter Cottontail from eating your carrots and even find out what plants would grow best in your area.

Suppose you want to create a backyard pond where frogs can frolic in the lily pads. You'll need to find out both how to create the pond and how to start the vegetation in it. A good place to begin would be:

```
gardening +planning pond
```

Other Simple Searches can also provide you with focused garden information. Try:

```
+gardening +shows "flower and garden"
```

for information on flower and garden shows, or:

```
+gardening +hydroponic
```

for specific information on hydroponic gardening, such as tips, CD-ROMs, supplies, and mailing lists.

genealogy

For starters, you can search for your family name and the word *genealogy* to see if someone on the Web is systematically recording information about your family. Do this by entering a Simple Search such as:

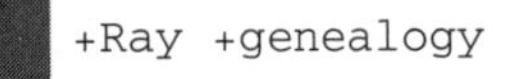

```
+Ray +genealogy
```

Or, if family tradition indicates that you might be descended from a historical figure, such as Pocahontas, you might want to track down and verify the connection. All you have to do is a Simple Search of the Web:

```
+Pocahontas +genealogy
```

The results of this search give you a good starting place.

In Search of Genealogy

In a comment about AltaVista Search, George wrote:

> *My main interest in the Internet is genealogy. I use AltaVista on a daily basis to scan for the surnames I am researching. I set the Advanced page to search the Usenet and then look for "genealogy and xxxx" substituting a surname for xxxx. It works well for all my surnames except Price. That one comes up with too many false hits to use.*

—George B. Whaley

geography

Suppose you're trying to do a research paper on a geography-related topic but can't decide on a specific angle. Let AltaVista Search help narrow your ideas! For starters, you can do a broad search, such as:

```
geography
```

This search provides you with data on a huge range of geography-related topics, such as physical and cultural geography, geographical anomalies, and historical geographical sites. Now suppose you've narrowed your topic to the geography of your favorite obscure region. You can take a look at this area by searching for Internet map viewers, like this:

```
geography map viewer
```

or:

```
PARC Map Viewer
```

Either of these searches provides links to online geographic maps that allow you to zoom in on specified regions, cities, or towns. You can even access the information that was used to create the maps by searching for:

```
GIS geographic information system
```

Also, you can search Usenet for:

```
gis faq
```

and find out how these cool maps work or search for:

```
geoscience gis faq
```

for related geographical information.

getting started

With what, you ask? Almost anything. Try a Usenet search for:

```
"getting started"
```

and just start browsing. You'll find tips and tricks oriented to the beginner in almost every subject under the sun. If you have specific interests, however, feel free to add them, as in:

```
"getting started" showing GSD
```

for a world of tips about showing your German Shepherd. (GSD stands for German Shepherd dog—all of those in the know use the acronym.)

Or suppose you're really into aviation. Check into a Usenet search for:

```
newsgroups:aviation.homebuilt getting started
```

for tons of information on—what else?—getting started with building your own aircraft and related paraphernalia.

government

If you're interested in getting all you can out of your taxes, be sure to take advantage of the government information available on the Internet. Just a quick search for:

```
government
```

produces a number of links to information. If you need specific information and, for whatever reason, particularly want governmental sources, you can always try a domain search of the Web. For example, to find information about taxes (ugh!) from only government sources, you'd search for:

```
domain:gov taxes
```

On the other hand, if you want to see some of the cool things that the Library of Congress is doing, just search for:

```
host:loc.gov exhibit
```

We assure you, it's almost as good as a museum. Speaking of which, a search for:

```
Smithsonian Institution
```

gives you a whole list of the online presentations that the Smithsonian makes available, among other information.

GPS

Technology at its best! Using the Global Positioning System (GPS), you can find out the exact location of anything on earth—really! For example, try a search for:

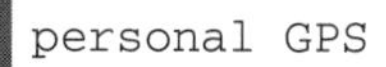

```
personal GPS
```

to find out exactly where you are. Or, if you're just interested in information about GPS and its uses, try a search for:

```
personal +GPS using stor* experience
```

Here, you'll find explanations, anecdotes, and uses for this cool technology.

H

hand-held computers

If computers are so much fun that you have to take one everywhere and you think that even the newest notebook computers are far too big and clunky to use effectively, try:

```
hand-held computers
```

Really! This search gives you loads of information on this technology. Or, if you're looking for specific models, try something like:

```
hand-held computers +newton
```

Over on the Usenet side, you'll probably want to get some input from other users before you take the plunge into hand-held computing.

In Search of Health

In a comment about AltaVista Search, Kala wrote:

> *I teach health policy to graduate students. I currently have a class of nine with mostly clinical backgrounds, several with little or no computing experience. I put them online and showed them how to access AltaVista (you are at the top of a page I have organized for them to use in doing health policy research). Within ten minutes each had completed and refined a search for their paper topic. The ease and intuitiveness of the system was clearly motivating even the ones with no prior experience (and anyone with none today has had some phobia about working online). They were all excited and enthusiastic and sold on the capacity to do research online. Thank you for a terrific teaching tool.*

—Kala Ladenheim

herbs

Herbs? No, not your Uncle Herb (although you could check out "acquaintances" or "genealogy" in this Reference to do research on him). You can look up *herbs*, those tasty weed-looking things you add to your food. For general information, enter:

```
herbs
```

which gives you general information about different kinds of herbs as well as cooking with them and growing them. If you want to narrow your search, you can enter one of these queries:

```
herbs growing
herbs recipes
herb* medicine
```

These searches provide you with information on herbs as they pertain to the specific use you're looking for. If you're interested in the latest discussions and opinions, try a Usenet search for:

```
herb* medicine
```

Also, take a look at the search ideas in the "cooking" section of this Reference.

hexadecimal

What the hex is that?

Don't you get tired of seeing "hexadecimal" thrown around like it made sense? Just check with the Web and you'll get the real meaning. A search like:

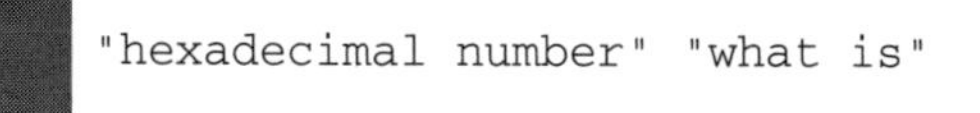

```
"hexadecimal number" "what is"
```

brings up several pages with definitions and explanations.

hiccups

Have you ever wondered what causes hiccups? Now you can find out! If you do a Simple Search for:

```
hiccups
```

you'll find all sorts of information on the causes and some recommended cures for those racking hiccups. You can even find specific information on, for example, those tiny hiccups an unborn baby makes by narrowing your search, like this:

```
+hiccups +prenatal
```

hobby

What do people do with their free time? Perhaps more important, what should you do in your spare time? Try a Simple Search for:

```
hobby
```

and see the variety of crafts and activities available. Whether you have spare time on the weekends, just a few minutes here and there, or a few weeks to spend relaxing, these results will give you some nifty ideas.

Or, if you have a specific hobby in mind, type it in, like these examples:

```
aircraft spotting

bird watching

stamp collecting

dog training
```

home buying

Suppose you're in the market for a new home and need to learn more about it. A quick Simple Search for:

```
house buying
```

yields tons of pages from realtors and others in the real estate industry who will be happy to advise you. Even realtors out of your area can be helpful—most tips aren't regional in nature. (By the way, don't bother searching for *home buying* because "home" is so common on the Web that you'll get nothing worthwhile.)

While you're at it, check out:

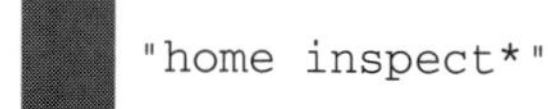

```
"home inspect*"
```

to find links to information about home inspectors—always a good idea. Then, toss in your city name when you're ready to move from getting broad information to picking someone to actually do the work.

home repair—electrical

Just can't figure out what to do about that persistent problem with fuses blowing? Try a Simple Search for:

```
fuse blow*
```

If you're concerned about the fuses and your household wiring in particular, you could always add:

```
fuse blow* +house wiring
```

or:

```
"house wiring" +"circuit breaker"
```

If everything works, but you just can't figure out why the light switch gets so hot, check into:

```
"fire code"
```

or:

```
"fire code" +electrical
```

ıome repair—plumbing

On the other hand, maybe you're tired of paying the big bucks for a plumber to come and take care of what looks like an easy repair. (It never fails that the plumbing backs up on a Sunday afternoon when you have six relatives visiting, right?) Do it yourself! Just search for:

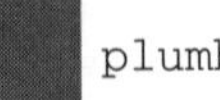

```
plumbing instructions
```

One tip: Before you start, do a quick search for:

```
plumbing contractor
```

or:

```
"repair* water damage"
```

in case something goes wrong.

> **In Search of Homework**
>
> In a comment about AltaVista Search, Cathy wrote:
>
> *I had to write to say that this site is a terrific Internet tool. Tonight my son (who lives in Brookings, OR) called me to say his junior high math teacher had asked them to find out what "sociable numbers" are. None of my limited reference books had the term, and in Sacramento we have very limited library services. So I looked it up here, and we were able to discover what the term means and what an aliquot cycle is. I would have never found it without you. And having found it, I was able to bring current research from the U. Conn. to a youngster living at the edge of nowhere. Thanks a million!*
>
> —Cathy Horiuchi

horses

Find happy trails with AltaVista in a search for information about horses. If you need some information before diving into equine ownership, try searching for:

```
horse
```

on the Web to find everything from horse museums to information about specific breeds and horse care. Alternatively:

```
equine
```

produces more scientific or medical links, such as *equine reproduction* or *equine clinic.* Usenet searches for:

```
horse
```

or:

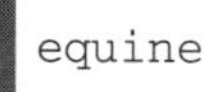

```
equine
```


produce thousands of links, mostly about horse health. If you just want to find horse discussions, try a search for:

```
newsgroups:equestrian
```

You might find other search ideas under the "zoo" section of this Reference. Check it out!

hovercraft

Sure enough, even "hovercraft" have made it to the Web. Just a search for:

```
hovercraft
```

gives you links to pictures, schedules, and hovercraft for sale.

humor

A great way to find the funny of the day is to search in Usenet. For a more or less complete rundown on the latest (usually heavily recycled) humor, search for:

```
newsgroups:rec.humor -newsgroups:rec.humor*.d
```

to get all the jokes (but not the discussion in the .d groups) from the humor group. You can also search for:

```
humor -tasteless
```

if you want to avoid the…er, um…*tasteless* jokes out there. (Actually, the search excludes only the messages clearly labeled as such, but it's a start.) Unfortunately, there's no way to find only the truly hilarious jokes, but:

```
newsgroups:rec.humor.funny
```

usually has some good material. See for yourself!

On the Web, thousands of sites claim to provide humor. Try searching for specific kinds such as:

```
humor limericks
```

I

igloos

You'll probably have to print the information and take it with you, but be prepared with instructions for winter wilderness survival. Try a Simple Web Search on:

```
igloo building
```

to find out how you can build an igloo of your own. Or to seek out a broader perspective on the topic, try:

```
"native american" living indian* igloo eskimo
```

This search provides links to information on why and how igloos became part of Inuit living. If, on the other hand, you're more interested in the wilderness aspect than the Inuit angle, try a Usenet search for:

```
newsgroups:backcountry survival
```

In Search of Ice Cores

In response to a discussion of religion and the age of the earth, Kevin wrote:

> *There have been projects in Greenland and the Antarctic where the layers are particularly thick.*
>
> *But as for specifics, you might go to altavista.digital.com and search for +Greenland +"ice core." You'll get about 400 hits. Near the top of the list is "GRIP Greenland Ice Core Data, Paleoclimate Data; NGDC (EARTH_LAND_NGDC_PALEO_GRIP1)" which says that a particular ice core taken in 1991 was 3,028.8 km in length and goes back 250,000 years. I guess if there were a big layer of dirt from the Flood of Noah, they would have mentioned it.*
>
> *If you add +Noah to your search string, then you will skip the pure science and get into the polemics.*

—Kevin Davidson

9

indoor plants

Are your houseplants dying faster than you can buy new ones? Wondering how to get that green thumb you've always wanted? Do a quick Web search for:

```
houseplants
```

This search provides you with tons of information about indoor plants, including tips on caring for plants, information on which types will grow best in your household, and ideas on purchasing plants. You can even find out why your favorite plant seems to be withering away. Just do a Usenet search for:

```
+plants dead dying
```

If you know enough about the plant to be more specific, by all means do so. For example, a Usenet search for:

```
jade plant dropping leaves
```

works wonders, if you have a jade plant that's, well, dropping leaves. If you just have to have plants but all other means of keeping them alive fail, try a different type:

```
+cactus +plant newsgroups:garden
```

Cultivate your knowledge about cactus plants (we hear they're practically impossible to kill!).

See also the "gardening" section of this Reference for more ideas on plant-related searches.

investments

Thinking about investing? Not sure you're ready to call a broker? Try doing some research on the Internet. A Usenet search can provide you with advice and answers to questions. Just enter:

```
+investment advice +newsgroups:invest*
```

If you don't want to hear the online sales pitches, you can try:

```
+investment advice +newsgroups:invest* -guaran*
```

which weeds out the "guaranteed" investments. You can also check out the Web for investment information by typing:

```
+investment advice
```

This search yields a number of scholarly articles and apparently sound information.

J

jargon

The world of computers is filled with jargon, most of it remarkably arcane and hard to decipher, including terms like *SCSI, hacker* or *cracker, kluge,* and *abend,* to name only a few. To find the meaning of these and practically any other jargon term, search the Web for:

```
jargon dictionary
```

Not only can you go out and buy a dictionary of jargon, you can also search out the meanings right on the Web with a couple of quick clicks. That is, if someone calls you a "suit" in an apparently derogatory way, do a search for:

```
+jargon +suit
```

jazz

Now here's something to get jazzed about! All that jazz, ready to be seen, heard, and enjoyed, awaits at:

```
+jazz festival
```

With this search, you'll find all sorts of links to information about jazz festivals, their history, and their growing popularity. If you'd rather experience intimate improv sessions, or even get into making the music yourself, try:

```
+jazz improv*
```

If you can't play the music without an instrument, get one through a Web search for:

```
music instrument +"for sale"
```

Over on the Usenet side, you'd be well advised to include the name of the instrument (unless you're pretty multitalented), like this:

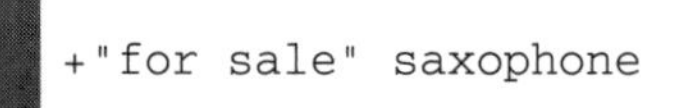

```
+"for sale" saxophone
```

should get you anything you need, except perhaps a booking on Leno:

```
+"tonight show" +"Jay Leno"
```

jobs

The Internet is *the* place to look for job postings. If you're just browsing (and want to browse through a lot!), try a Simple Usenet Search for:

```
newsgroups:jobs
```

This search results in lots of listings, probably more than you can wade through, so you might focus your search slightly, for example, like this:

```
newsgroups:misc.jobs
```

The misc.jobs newsgroup is where many postings are found. If you have a local set of newsgroups, such as ba. for the Bay Area, or ok. for Oklahoma, you might head for Advanced Search so you can add that:

```
newsgroups:(ok.jobs or ba.jobs)
```

Of course, if you're looking for a specific sort of job, you could add that in the ranking area of the Advanced Search, but you shouldn't be too specific.

If you're a writer, you'd put:

```
writer
```

in the Ranking field. And there you go! All the jobs you'd ever want.

Using AltaVista to search the Web can also help you find specific jobs at specific companies. Suppose you're interested in a job as a marketing assistant at a company called Symmetrix, a management consulting firm in Lexington, Massachusetts. First, try searching for:

```
Symmetrix
```

which gives you hundreds of hits, many of which deal with computer systems with that name. Since you're looking for a specific company by that name, try:


```
Symmetrix "management consulting"
```

and you get the site of the company you want.

If you knew your potential employer's name, you could search for it on the Web and in newsgroups to find out his or her background and interests. Doing so before the big interview could give you an edge over other applicants.

K

keys/locks

Just because you locked your keys in the car doesn't mean that you're out of luck. AltaVista Search to the rescue! Just search for:

```
lock key
```

or:

```
locksmith
```

These searches provide you with links to locksmiths who have Web sites. You'll probably save some time if you also add your location, like this:

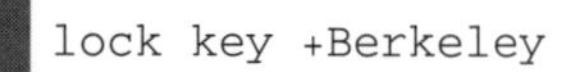

```
lock key +Berkeley
```

Heck, if you hurry, you'll be able to get the doors unlocked so you can get in and put the top up! By the way, if you make a habit of this (or if you have kids who play with the bathroom locks), you might search for:

```
lockpick* kids
```

In Search of a Kid's Story

In a comment about AltaVista Search, Don wrote:

> *I was showing one of our recalcitrant users just how broad the Web info is when we decided to search for something truly unusual and quintessentially Australian, so we fed in "Mulga Bill's Bicycle," the name of a reasonably well-known kid's story from many years ago. We got 15 hits in the blink of an eye. Most impressive.*
>
> *—Don Ewart*

kites

Oh, go fly a kite! What? You don't have one, and you say the tree down the street always takes the kite for lunch anyway? Turn to the experts in a Usenet search for:

```
kites
```

If you have specific issues to discuss, such as lights for kites (really!), try:

```
newsgroups:kites lights
```

If the weather's bad, just search over on the Web for:

```
+indoor +kite instructions building
```

knitting

When you've just got to find something to occupy your hands while you're waiting for those Web pages to download, do what German university students do in class to pass the time—knit! For starters, try entering:

```
knitting
```


This would give you information not only about newsletters and shows, but also about discussion forums where you can share tips and ideas.

Check out other knitting search ideas under the "blind" section of this Reference. Cool stuff!

L

lace-making

Although you might have thought that lace came only on big spools in the fabric store, you'll be pleased to learn that you can make your own, and the Web can tell you how. Just use AltaVista and search for:

```
lace*making
```

This Simple Search finds a surprisingly long list of resources, ranging from materials to actual how-tos. Searching for:

```
lace
```

gets into a whole different world—something about "leather and…."

languages

Just can't go on without brushing up on your language knowledge? Try:

```
language learning
```

or:

```
language acquisition
```

to get the latest information about how people learn languages. Throwing in Noam, as in:

```
language acquisition +Chomsky
```

gives you the whole black box theory, online. To narrow it down a little more, try:

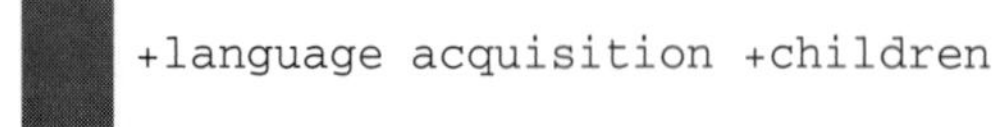

```
+language acquisition +children
```

and see how your kids (or you, as a kid) did it. Also see the "foreign language and culture" section of this Reference for more neat language search ideas.

legal research

While AltaVista isn't a good substitute for a lawyer, you should certainly use it to get a head start. For example, if you're buying a house, you might try a search like:

```
sample +"real estate" contract
```

to get an idea of what issues you should consider when making an offer on a house. Or suppose you want to get into breeding the animal of your choice. Simply check out:

```
sample stud contract
```

If you're looking for general information, remember to use wildcards in your search, like this:

```
legal form*
```

to get sample contracts and forms for a number of purposes.

Of course, you'll want a real lawyer to check out the contract before you use it for anything but entertainment, so try Business Search, choose Attorneys, then add your city and state. Click Search and choose from the list. Add an attorney name if you like.

Finally, all of this legal stuff is likely to get tiresome quickly, so get some entertainment out of it as well. Try searching for:


```
lawyer jokes
```

lesson plans

If you're a teacher and completely out of ideas for that lesson coming up, try searching the Web for:

```
"lesson plan*"
```

You'll find hundreds of prepackaged lesson plans, ready for use. Many of these have been successfully used by teachers around the world! Add the subject or area for a more precise result, as in:

```
"lesson plan*" +german
"lesson plan*" +science

"lesson plan*" +health
```

Ahhh! To be in fifth grade again, making string art, designing rock gardens, and coloring purple people eaters.... Those were the days.

In Search of Librarians

In a comment about AltaVista Search, Allison wrote:

> *I am so glad I happened on your site. Your search engine is by far the slickest I have used. As a librarian, I appreciate the scope of information to which you provide access. I found a picture of Diana Ross, the dimensions of the Statue of Liberty, and the names and addresses of shops selling brass compasses all in one short session. I am telling others about your service. Thanks again.*

—Allison Williams

lighthouses

Lighthouses, too, find a home on the Web. A search for:

```
lighthouse picture
```

produces links to pictures of lighthouses around the world. If you're more into the magic and romance than just seeing a picture, try:

```
lighthouse cookbook
```

to eat as the lighthouse keepers did, or:

```
lighthouse book
```

to read the stories of how lighthouses were built as well as how they were used and maintained. If you're interested in buying a lighthouse, a search like:

```
+lighthouse sell buy sale real estate
```

should produce any leads available on the Web.

> **In Search of Logan's Run**
>
> In reference to a question about the movie *Logan's Run,* Cassie wrote:
>
> *I did a Simple Search in AltaVista and found the FAQ here (first try!). If you ever want to try AltaVista (great search engine), the URL is:*
>
> *http://altavista.digital.com/*
>
> —Cassie Chamberlain

lyrics

AltaVista Search can open the door to the lyrics of thousands of songs. Try a search for the title of your favorite, or just head for lyrics archives to choose from thousands. A search for:

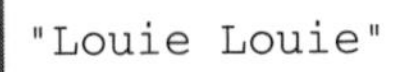

```
"Louie Louie"
```

yields hundreds of "definitive" versions of the lyrics to this frat party standard.

Take a look at the "music" section of this Reference for more information about searching for song titles and albums.

M

marketing research

Suppose you are a supplier of Internet products and services. You do business mainly through distributors, so there are many customers using your products that you have never dealt with directly. Many of those may choose to display your company's logo at their site. Those that do will probably keep the original filename for the .GIF file of your logo. If you work for Digital Equipment, you might search:

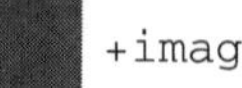

```
+image:digital.gif -url:digital.com
```

because you are only interested in finding non-Digital sites. You get about 5000 hits, many of which in fact do display Digital's logo. (Others simply have given the name DIGITAL.GIF to other graphics files.) The name *AltaVista* is more distinctive and rare, and a search for:

```
+image:altavist*.gif -url:digital.com
```

yields about 800 matches. (You search for *altavist** rather than just *altavista* because some systems are limited to filenames of eight characters.) In this case, it looks like most of the matches are good ones.

You might be tempted to use the same technique to do research on banner advertising on the Web. At first it would seem that a search for the filename of the .GIF used for a banner would match all sites now running that ad. But many (probably most) Web sites with paid advertising rotate which banner appears automatically. So it would be random luck whether the .GIF you are looking for happened to be displayed when AltaVista Search visited to index the page.

In Search of Mars

In reference to a question linking Carl Sagan to faces found on Mars, Dave wrote:

> *I think Carl Sagan would rather jump through a burning ring of fire than have his name associated with the Mars Face other than in a debunking manner.*
>
> ...
>
> *That last URL [omitted] is for an AltaVista search result which will give you a bazillion pro/con/whatever references on the subject. Certainly, more references than I could list here.*
>
> *By the way, I highly recommend AltaVista [http://www.altavista.digital.com] for any serious Web or Usenet searches. There are a lot of search engines out there, for a lot of different purposes, but for the Web and the Usenet you're not going to find an engine with the depth and up-to-date qualities of AltaVista. If you read the search instructions carefully so that you know how to get around, you'll be amazed at the results it gives.*
>
> —Dave Locke

medical topics—general

AltaVista Search lets you find a vast range of information on medical topics. Of course, AltaVista Search *cannot* take the place of a visit or phone call to your doctor, but it can provide you with a good place to start researching.

AltaVista Search is a great way to glean medical information and anecdotal data. Start by searching for specific conditions—the more specific the better. For example, suppose you've been experiencing lower back pain. You can enter the Simple Search:

```
"lower back pain"
```

This search gives you a wide variety of information on the causes and remedies for common lower back problems. If you've noticed other symptoms that accompany the back pain, you can enter those as well:

```
"lower back pain" "leg"
```

This more specific search narrows the results to just those that apply to the symptoms you entered.

Or suppose you just came home from the doctor with a prescription and want to know what kinds of side effects to expect. Instead of dragging out that huge *Physicians Desk Reference* (PDR), just look it up using AltaVista Search. For example, for information on a specific drug, just enter the name, like this:

```
"Tylenol 3"
```

medical topics—rare

Normally, the rarer a medical condition, the more difficult it is to find information about it or to get in touch with others who have experienced it. With AltaVista Search, rarity is an advantage, helping you to get to what you want quickly. For instance, a Simple Search of the Web for:

```
"Opitz Syndrome"
```

yields a list of a couple of dozen pages, all of which are on target, including explanations of the condition as well as parent support groups. Likewise, a Simple Search of the Web for:

```
"sensory integrative dysfunction"
```

provides a couple of dozen useful matches.

medical topics—specific

Suppose your child, born with cleft lip/cleft palate, is now a teenager and is scheduled for cosmetic nose surgery (rhinoplasty). You'd like to learn whatever you can about this procedure—especially in cases involving cleft lip. First you try a Simple Search of the Web:

```
+"cleft lip" +rhinoplasty
```

That yields dozens of matches, many of which are lists of the capabilities of medical institutions that do reconstructive plastic surgery. You'd prefer information that specifically links the condition of cleft lip with the procedure of rhinoplasty. So try Advanced Search and use the *near* operator, like this:

```
"cleft lip" NEAR rhinoplasty
```

This yields useful information on rhinoplasty, but still with no connection to cleft lip. So you'd like to check for parent support groups, some of whose members' children might have gone through a similar operation.

```
+"cleft lip" +parent* support
```

helps you locate Prescription Parents and other such organizations.

In Search of Medical Information

In a note about AltaVista Search, Jim and Alison wrote:

> *I just thought I'd send a note of appreciation to you as I've just used your page to search for information on chicken pox, which my five-year-old son has contracted. The amount of info forwarded by your page was amazing, and has turned us into the "world's most knowledgeable" on the subject!*

—Jim and Alison Hanner

meeting planning

You've just been nominated to plan your company's annual board meeting, which will take place in San Diego. Gads! You need to find a hotel, meeting rooms, transportation—the whole shebang. You can start by searching the Web, like this:

```
+hotel + "San Diego"
```

You get about 200 matches, many of which look like they are right on target. From there, you can narrow the search to a specific hotel and search for meeting rooms and prices, as in:

```
+"San Diego" +Ramada "meeting rooms"
```

Many organizations, such as car rental companies, provide prices—or at least contact information—on their Web sites. So, if you're looking for prices on car rental, just enter the *company name* plus *price,* like this:

```
Avis price
```

Check out the section on "resorts/convention centers" in this Reference for more ideas.

metal

Let's say as part of your job, you have to be on the lookout for suppliers of sheet metal for backplanes, preferably ones located in Massachusetts. Try Advanced Search of the Web, like this:

```
("sheet metal" AND backplane) AND (Massachusetts OR Mass OR MA)
```

Or suppose you inherited gold coins and want to know their current value. Try entering the search:

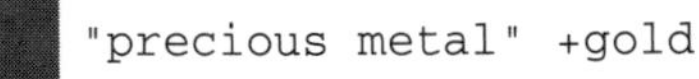

```
"precious metal" +gold
```

This search gives you records of the price of gold and links to other information about gold.

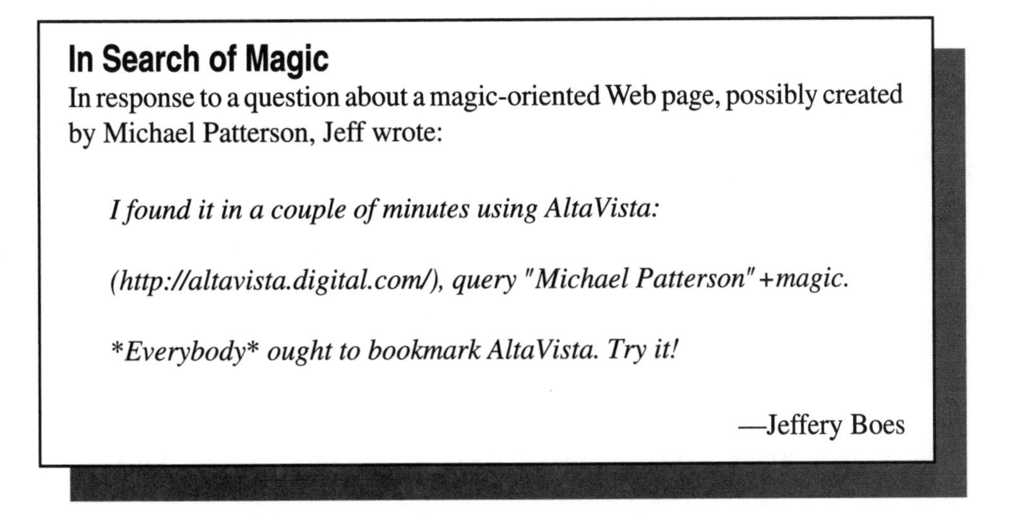

In Search of Magic

In response to a question about a magic-oriented Web page, possibly created by Michael Patterson, Jeff wrote:

I found it in a couple of minutes using AltaVista:

(http://altavista.digital.com/), query "Michael Patterson" +magic.

**Everybody* ought to bookmark AltaVista. Try it!*

—Jeffery Boes

money

To see if money is just waiting for you to pick up, do a Simple Search for:

```
"unclaimed refunds"
```

No kidding! An increasing number of states and institutions post on the Web lists of tax and other refunds that individuals have not yet collected. Maybe there's an unclaimed refund with your name on it.

Also, try an Advanced Search for:

```
(unclaimed AND bequest*) OR (missing AND heir*)
```

This yields about 14,000 matches, many of which are references to TV shows with those elements in the plot. So refine the search, like this:

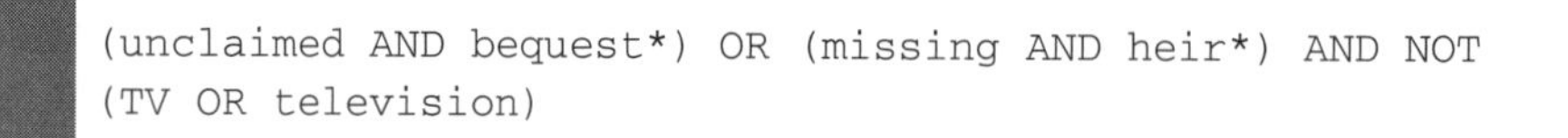

```
(unclaimed AND bequest*) OR (missing AND heir*) AND NOT
(TV OR television)
```

That narrows the field to 10,000. But if you seriously think that there might be money coming to you, it might be well worth your while to enter other relevant words or even check them all.

movies

Just have to have more information about a specific movie? Can't remember who played which role? Just search for the movie by name in a Simple Search of the Web, as in:

```
Braveheart
```

Need to find a good—no, great—flick for weekend viewing? Review pages about the winners to help you decide:

```
oscar winner "best picture"
```

On the other hand, if you're just looking for a quick film clip to play on your computer to verify that it really is the powerful beast the salesman assured you it was, search for:

```
movie archives download +trailer
```

In Search of Movies

In reference to a movie allegedly based on the old role-playing game Zork, Kathleen wrote:

> *I did an AltaVista search for zork movie and found a page that said Activation and Threshold Media had agreed to make a movie and TV series based on the Zork game.*
>
> —Kathleen S. MacPherson

music

AltaVista Search plays to the tune of thousands of music topics. Check these out, just for a demo. Suppose you'd like to have some lyrics of French songs. All you have to do is search the Web by entering *song,* like this:

```
song
```

This search gives you thousands of hits…er, um…song titles, as well as information on songs and music in general and even a country song-naming contest. You might try narrowing the search by entering words from the song title or the album title, like this:

```
"I'm just a bill"
```

or:

```
"schoolhouse rock"
```

These searches give you, among other things, a link to the Schoolhouse Rock series of kids' music. (You remember these, don't you? "I'm just a bill, yes I'm only a bill, and I'm sittin' here on Capitol Hill...".)

AltaVista Search can even help you book your band for gigs. Try searching newsgroups for:

```
+guitar +gig* +wanted
```

With this search, you can find all sorts of organizations advertising for entertainment help. Just what you're looking for!

Along those same lines, suppose you are interested in a band, but aren't sure how the name is spelled. You think that it's "Verucha Salt" or "Berucha Salt" but aren't sure. Searches on those phrases yield nothing. So you search for the name of one of their songs, "Seether," on the Web, which immediately gives you a long list of pages devoted to "Veruca Salt." Now, knowing the right spelling, you search for the group's name and get over 1000 good matches. Following those links, you might be tempted to buy a CD at one of the commercial Web sites. For instance, "American Thighs" sells for about $16 (including shipping). But before placing your order, you should search for the same CD in the newsgroups:

```
+"Veruca Salt" +"for sale" +"American Thighs"
```

where some individuals are offering it for sale (in "mint" condition) for $7 or $8.

In Search of Music

In reference to a question about an October Project CD, Chad wrote:

> *They are more of a college alternative band, actually (although some of the elements of this band might appeal to Enya/Lorenna McKennit fans)... I believe they are just 5 people from a particular college (forget which one) that formed a band and got discovered. There are about 4 or 5 Web pages on them, which searching AltaVista (http://altavista.digital.com/) for october+project will find.*

—Chad Gould

N

names

If the stork's arrival is any day now but you still don't have a good name picked out, do a Simple Search for:

```
selecting choosing "baby name*"
```

Everything from lists to programs that provide choices for baby names will appear. On the other hand, if your interest in names is a little different, search for:

```
"hurricane names"
```

If you want, you can ensure that your child's name will be the first hurricane of the 1999 hurricane season in the Atlantic Ocean (that is, Arlene).

name that publication (or Web site)

With the vast numbers of Web sites and online publications, it is becoming increasingly difficult to think up a unique name for a new one. AltaVista Search helps you quickly test your ideas, to see if someone else is already using that same name. For example, try a Simple Search of the Web for:

```
"Internet to go"
```

There are too many matches to try them all. But a quick check determines that many are just random appearances of those three words in order, such as "...Internet. To go...." Try *Internet to Go,* capitalizing "Go" as more likely in the name of a publication or Web site. As it turns out, several companies have already announced a product with that name. If what you plan to do does not in any way compete with that product and couldn't be confused with it, you might consider using a typographical variant, such as *Internet2Go*, which yields (as of now) zero matches.

NASCAR races

Just can't get to the TV for the latest on the NASCAR races? Do a Web search for:

```
NASCAR
```

(yes, use all caps) to keep up with the results and follow along with driver profiles.

natural history

If you're losing sleep because you just don't know how old the earth is, check with AltaVista Search. A Simple Search for:

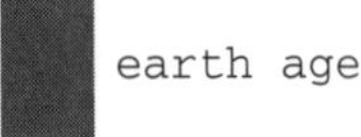

```
earth age
```

should get you going with geological, biblical, and other interpretations of the age of the earth.

New Age

Suppose you're researching New Age religions or even thinking about joining one. AltaVista Search can provide you with scads of information. For example, check out a Usenet search, like this:

```
newsgroups:newage
```

This search brings up a variety of discussions and opinions about religion, generally in the context of New Age religion. You can also check out New Age music by searching Usenet using the phrase:

```
"new age music"
```

for discussions on this topic. Or try the same search on the Web to find everything from samples to description to reviews.

news archives

Even though the Web doesn't land in your bushes each morning, it's still a good source of news. Try a Web search for:

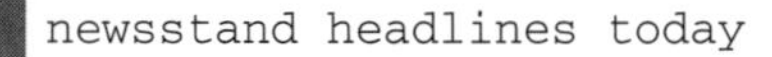

```
newsstand headlines today
```

You'll find links to a number of sites that provide news online. Although many of the sites provide only partial information for free and require a subscription for the whole story, it's a quick and easy way to catch up on the headlines.

O

opinions

Just the facts, ma'am, just the facts. Or actually, you can even find opinions, ma'am. Do a Usenet search, any Usenet search, and you'll get more opinions than you probably bargained for. Sometimes these differences of opinion result in "flame wars," in which heated opinions and insults are exchanged.

If you just want an interactive editorial page, try searching for the controversial topic of your choice, along with *newsgroups:talk.* For example, enter:

```
taxes newsgroups:talk
```

or:

```
newsgroups:alt.politics "government size"
```

Keep in mind, though, that these (and similar editorial pages) are just like the opinion page of your newspaper—you might not agree with the opinion.

In Search of Old Friends

Deb wrote:

I have a friend that I've kept in touch with for over twenty years—since we were in first grade together. Unfortunately, after I got married and moved a few times, he and I lost touch. I tried an AltaVista search for his first and last name; then I tried his first and last name plus the state in which he had last lived. I finally tried his first and last name, plus his profession, like this

"firstname lastname" +music

Interestingly, I found him through a music company's Web site that listed his latest CD. Outstanding!

—Deb Rowe

outdoor activities

Suppose you've had enough of the computer and Internet world and need to get away—really away. (It *could* happen, you know.) Before you hit the power switch and wander off into the great outdoors, try out a Usenet search on outdoor-related topics, for example:

```
newsgroups:backcountry survival skills .
```

You'll find useful information on surviving your adventure as well as tips to make the most of it. You can even find great debates to ponder, such as a discussion of cotton versus polyester sports gear.

Over on the Web, you find more substance and less talk with a search for:

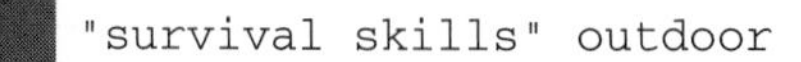

```
"survival skills" outdoor
```

After you've decided to tame your trip and not camp out with the bears, you might try a Simple Search for:

```
outdoor recreation camping backpacking
```

Have a good trip!

P

parenting

From traditional parenting to caring for elderly parents, a Web search for:

```
parenting
```

produces thousands of hits. With this search, you find links to the latest tips and trends as well as tried-and-true techniques. If you're looking for specific information, try narrowing your search, like this:

```
parenting infants
```

This Simple Search brings up resources specifically geared toward the wee ones, while:

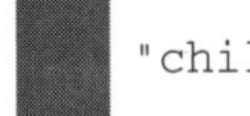

```
"child development" theory
```

provides sites that explain exactly what is going on every step of the way. You can also check out Usenet discussions, which provide perspectives that theory and textbooks rarely offer about raising children. You can find discussions on a variety of topics with searches like these:

```
child development
co-sleeping
breast-feeding
```

These Usenet discussions open up new sets of opinions and incredible varieties of information, from strictly scientific to personal views from all camps of child-rearing.

peanut butter

Just can't stand the thought of another peanut butter and jelly sandwich? You've heard of peanut butter and orange juice but it sounds pretty bad to you? Just search for:

```
peanut butter
```

and collect all kinds of recipes, including several enticing variants on chocolate peanut butter. If you're a peanut butter fan and tend to overdo, search for:

```
"peanut butter" +choking
```

Also check out the section on "cooking" as well as the user story on "recipes" in this Reference for even more search ideas.

personality profile

If you're interested in personality profiles (those silly tests that know more about the inner workings of your psyche than you'll ever admit, even to your spouse), do a Usenet or Web search for:

```
"personality profile"
```

Or you can choose a specific personality test, like:

```
myers-briggs
```

You can even take a Myers-Briggs test right on your very own computer. Try searching for:

```
myers-briggs taking
```

pest control

Got something bugging you? Look up pest control information on the Internet. For example, try:

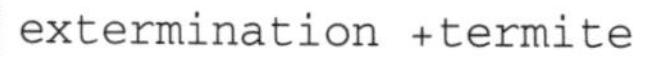

```
extermination +termite
```

This Simple Search yields broad information about pest control—everything from bugs to rodents to conferences you could attend. You'll also find specific information about exterminating termites and about doing so safely.

photography

Suppose you just got a great camera and want to know more about taking pictures. A general search for:

```
photograph*
```

yields all kinds of stuff—focusing, so to speak, would help. For example, if you want how-to information, try:

```
photograph* techniques
```

You can even find information on specific photographers, like:

```
photograph* "Ansel Adams"
```

As always, rare words help. For example, try:

```
daguerreotype
```

for links to information on old photographs, online reproductions, or even an article by Edgar Allan Poe about daguerreotypes. Or if you're looking for photos on a particular topic, enter:

```
pictur* photograph* +NASA
```

to get some space photos and the use policies on them. Or try:

```
pictur* +NASA
```

to go directly to space pictures. Really!

You can even keep up with the latest art photography trends by searching for:

```
lomography
```

poetry

"Whose woods these are I think I know...." If you *don't* know and want to find out more about this Robert Frost poem, let AltaVista Search help you out. Start with a Web search for:

```
"Robert Frost" woods
```

to find out more on this poem. Or:

```
newsgroups:rec.arts.poem*
```

will send lots of poetry information gently tapping, tapping at your chamber door. If you're not much into analysis, just search for:

```
poetry newsgroups:marketplace
```

to seek out used poetry books for sale.

politics

Yes, politics have even invaded the Internet. For the most timely information, do a Usenet search for:

```
"us politics"
```

This search provides you with all sorts of vehement opinions about every aspect of U.S. politics. You can also search for:

```
government conspiracy
```

which yields the world of entertaining reading you'd expect from a search on the word *paranoia.* Searching for your favorite political party on the Web would be fruitful. Just enter:

```
Republican
```

or:

```
Democrat
```

or even the name of your favorite candidate, like this:

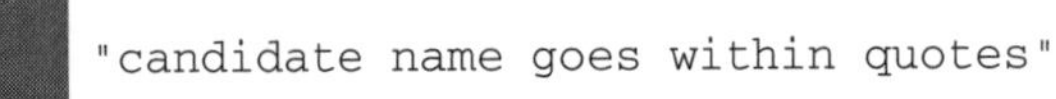

```
"candidate name goes within quotes"
```

You'll find more interesting information and heated discussion by searching for specific topics, such as:

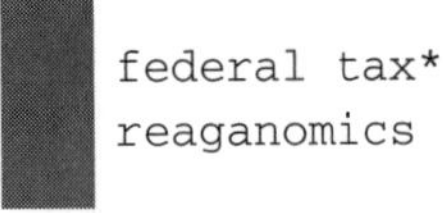

```
federal tax*
reaganomics
```

or:

```
immigration
```

publishing

Just about through with the Great American Novel? Need to find a publisher? Simple. Search for:

```
publisher contact information book fiction
```

or search by the publisher's name. If you get tired of sending your manuscripts to ungrateful publishers without the common sense to select your novel out of the hundreds they get each week, just look into:

```
self-publishing
```

and do it yourself.

Or suppose you are a publisher of children's books and believe that there is a market for fiction that deals with divorce and that involves pets. You'd also like to check what the competition has in print already, but since most online book catalogs are in databases, AltaVista Search would not have that information in its index. You'd also like to check the marketplace, looking for all references to such books on the Web. And ideally, you'd like to find works of that kind that are available on the Web that have not yet been published in print. You try a Simple Search of the Web:

```
+fiction + child* +divorce*
```

That gives you 1000 matches. Refining the search to look for unpublished works:

```
+fiction +child* +divorce* +unpublished
```

gives you about sixty to check out.

Q

quilting

If you or someone you know enjoys quilting, a quick AltaVista search for *quilt** works wonders. If you have something in particular in mind, you could always try:

```
quilt* +"Double Wedding Ring" +pattern
+quilting +lessons
```

Alternatively, you could branch out into:

```
+needle +crafts
+Arts +Crafts
```

In Search of Quilts

In reference to a discussion about quilt pattern ideas, Susan wrote:

> *There are a lot of X-files Web sites out there with great graphics for ideas—do a search with AltaVista or one of the other good search engines and do some surfing for quilt ideas. Sounds like fun.*
>
> —Susan Drudings

quotations

AltaVista Search is a great resource for checking out quotations, from the mundane to the clichéd. The old gag about "I'd rather have a bottle in front of me than a frontal lobotomy"—well, it's actually the title of an old country and western song. To find out more, type:

```
"rather have a bottle in front"
```

This kind of search is also good for identifying the clever sayings that you probably don't want on your Web page, because everyone else has already quoted them. For an enormous collection of diverse Web sites, try:

```
"knowledge is power"
```

In Search of Quotations

In reference to a questionable quotation, Morris wrote:

> *For several weeks, a co-worker has had on the dry-erase board beside his desk the incomplete quotation "_________ is the last refuge of the incompetent." It had become a game for others at the office to guess the missing word and author. No one really knew.*
>
> *I submitted the query in various forms to more than one search engine and found some interesting data, but not the authentic quote. From AltaVista I not only matched multiple good sites, but found a complete reference to Isaac Asimov's essays (the author). The missing word is "Violence."*
>
> —Morris Shaw

R

radio

You can talk to people around the world and you don't even need a telephone! A ham radio is just the trick. For starters, try a Web search for:

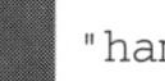

```
"ham radio" sale
```

and

```
+"ham radio" +regulations
```

for information about purchasing and using a ham radio. Also, you can find information about ham radios in general and the people who use them just by entering:

```
ham radio
```

In Search of Research

In a comment about AltaVista Search, Ian wrote:

> *I was doing research for a science fiction novel. I remembered from college, talk about a drug/medication/drink that caused your vision to change so that you seemed to be wearing yellow tinted glasses. This was an important fact in the novel. I typed in a command to AltaVista, something like +"distorted vision" +yellow +drug and within 2 seconds found a tea of sorts that is made from a plant that only grows in Mexico—the tea gives you a mild high like beer but if you take too much, you end up with yellow vision the following day. This worked perfectly into the outline for the novel and caused me to wonder what other source of information on the planet could have found that information in 2 seconds—what does this mean to man—what does it mean to potentially have every written word indexed in one place where it can be found in 2 seconds—it means tremendous progress for man and it also means that there is potentially the DEFINITIVE site—if it is not there—it doesn't exist.*

—Ian Ferguson

real estate

Planning to relocate? Let AltaVista Search help! Although it can't pack your boxes or lift your grand piano, it can find relocation assistance Web sites as well as cost-of-living calculators. For starters, try searching the Web for:

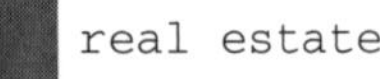

```
real estate
```

This search even finds real estate listings around the world! If you're just looking for a real estate agent, search for *real estate* and the location, as in:

```
real estate +portland
```

Also, you can find information about the place you want to move by entering *quality of life* plus the city, like this:

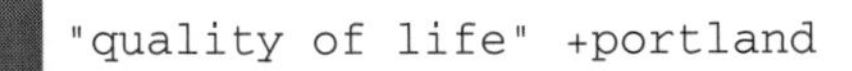

```
"quality of life" +portland
```

In Search of Recipes

In the context of a discussion of great Russian black bread, Kathleen wrote:

> *An AltaVista Search using the keywords "Russian Black Bread" yields many recipes. Try it!*
>
> —Kathleen Seidel

resorts/convention centers

Suppose your boss just put you in charge of finding a location for the company's next annual ball. You can use AltaVista to find convention centers for job or professional events (or for hobbies, we suppose). Try:

```
convention center +"Salt Lake"
```

to find information about tourism, convention centers, and hotel accommodations in the Salt Lake City, Utah area. Simply substitute the location of your choice, and use a + sign to ensure the location is included in the search results.

restaurant menus (local or far away)

Make your time on the road more enjoyable—scope out restaurants and menus before you ever leave home. For example, type in:

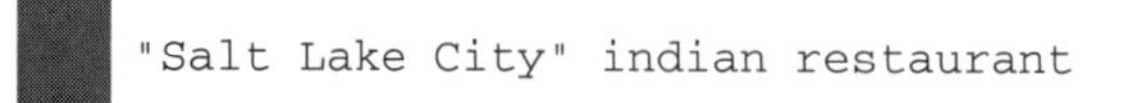

```
"Salt Lake City" indian restaurant
```

for a listing of Indian restaurants in the Salt Lake City area. You can also search menus to find interesting meals that you can cook at home. Simply check out a restaurant's menu, decide what sounds good, then search AltaVista and find recipes for that meal. Easy! See the "cooking" and "recipes" sections in this Reference for more ideas.

resumes

Whether you're writing a resume or just thinking about getting started, AltaVista is the place to look. For example, if you're writing a resume, try a Simple Search for:

```
resume
```

This easy search finds mostly resumes and links to resume listings that you can use to find examples of what other people are doing. If you're more interested in topics such as resume samples, design ideas, and tips, try narrowing it to:

```
writ* resume*
```

to get only resume writing tips and tricks. Or, if you just want models to go by, search for:

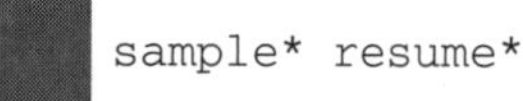

```
sample* resume*
```

roller coasters

Do you crave the thrill and excitement of roller coasters? Or maybe you'd rather read up on the roller coaster rather than actually having to hold on to your stomach for the whole trip? Try a Web search for:

```
roller coaster
```

or, if you're a traditionalist:

```
wooden roller coaster
```

to find out where to get to the old-time coasters. If you're more on the cautious side, there's always:

```
roller coaster safety
```

To discuss the physics of roller coasters, head to a Usenet search for:

```
newsgroups:roller-coaster physics
```

In Search of Academic Research

In a comment about AltaVista Search, Marty wrote:

> *I just thought I would let you know how much I appreciate AltaVista. I am working on my doctorate and your search tool is a lifesaver!!!! How fortunate we are to have your services in the Internet. I will forever be grateful to all of you. Believe me I will always tell everyone that AltaVista was instrumental in helping me complete my degree. Here is a big thanks!*
>
> —Marty Spitzer

S

scanners

So, you say you just bought a new scanner and want to scan every picture from your albums? No problem! Just try a Simple Search on the Web for:

```
+scanning tips FAQ
```

This search gives you all sorts of tips and tricks to making the most of your scanner and the scanned results. After you have that down (or possibly before), you'll need to do a search on:

```
"graphic* format" minimizing size
```

to learn about the different formats available and how to maintain quality without running completely out of disk space. Next, of course, you'll need information on:

```
disk compression
```

to try to squeeze as much space as you can out of the drive. Finally, the next step is to do a Web search for:

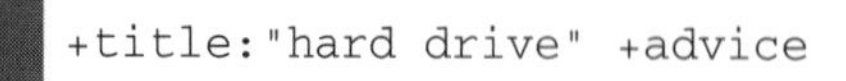

```
+title:"hard drive" +advice
```

to get information on more efficiently using your disk as well as on eventually purchasing a new drive. Of course, a Usenet search on:

```
buying "hard drive"
```

is also a useful place to start. See where buying that new scanner got you?

shareware

Shareware (try-it-before-you-buy-it software) is readily available on the Internet. To find it, just search the Web for the word *shareware* and your computer operating system, like this:

```
shareware Windows 95
shareware Macintosh
shareware OS/2
```

or:

```
shareware unix
```

You'll have your choice of dozens of shareware libraries. Do be sure and read the licensing agreements—much of the software isn't free, just free to evaluate. If you keep using it beyond a certain point, you must pay for it.

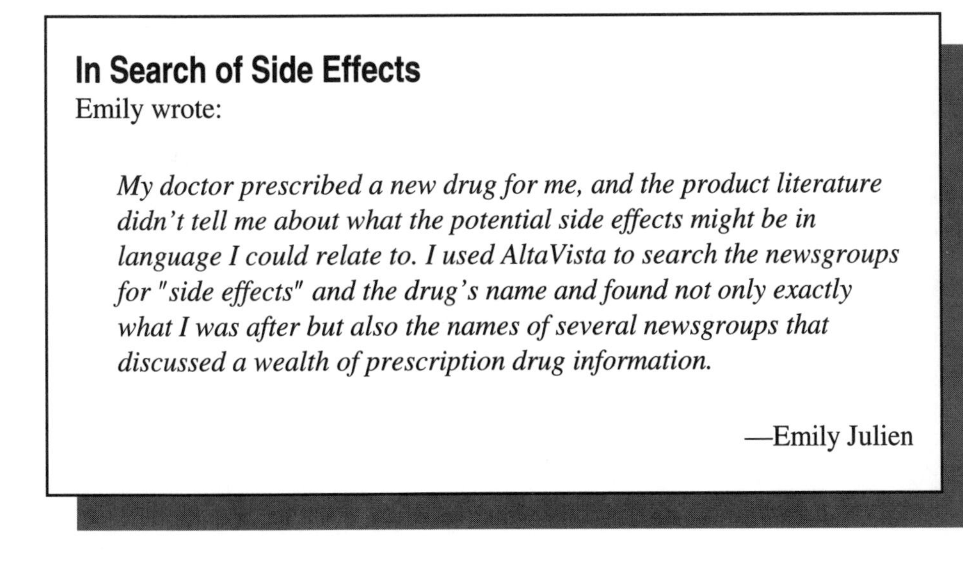

In Search of Side Effects

Emily wrote:

> *My doctor prescribed a new drug for me, and the product literature didn't tell me about what the potential side effects might be in language I could relate to. I used AltaVista to search the newsgroups for "side effects" and the drug's name and found not only exactly what I was after but also the names of several newsgroups that discussed a wealth of prescription drug information.*

—Emily Julien

skating

If you're tired of sitting at the computer, get your gear and head out to go skating. Do a Simple Search (Web or Usenet) for:

```
skating rink
```

if you're not sure where to go. If you need some help, try a Web search for:

```
"learning to skate"
```

for tips you can give your little one (or yourself!). You might also need:

```
"skating equipment" buy* sale rent
```

or even:

```
+"first aid" +skating +injury
```

(Read up on Colles' fracture!)

software

If you run your own company, say, as an Internet service provider (ISP), you may well want to locate billing software tailored for your needs. A Simple Search of the Web for:

```
+"billing software" +ISP
```

leads to a wide variety of packages you can choose from.

souvenirs

Just have to get the T-shirt, even if you haven't been there or done that? Not only will:

```
souvenir
```

find all kinds of sites about souvenirs, but using the site and a good credit card, you can buy them too. If you'd ever longed for an authentic Manx cat T-shirt, direct from the Isle of Man, you're set now! Search for:

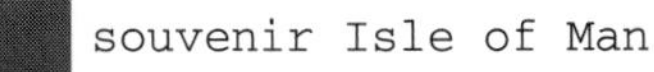

```
souvenir Isle of Man
```

If you don't care much about the souvenir as long as you get to take a virtual tour, there's always the Magic Kingdom, accessible through a Simple Web search for:

```
Disney
```

spam

It's not just for breakfast anymore! Actually, it's become one of the biggest Internet headaches of the year. What is it? It's the Unsolicited Commercial Email (UCE) that clogs up your email box every day. Want to find out more? Just do a quick search for:

```
spam UCE
```

Or you can find out about index spam (deceptive or bogus submissions to Internet search services) by searching for:

```
"index spam"
```

spelling

Suppose you aren't sure whether you should use "Webmaster" or "Web master." The term is too recent to appear in the dictionaries at your disposal. In Advanced Search, you select "As a count only" (instead of "Standard") in the choice of how to display results, and you check the Web for:

```
"web master"
```

and get a score of over 80,000 documents matching the query. Then you try:

```
webmaster
```

and get over 5,000,000. While these numbers are approximate, not precise, you find out that the overwhelming majority of people on the Web spell it as one rather than two words.

sports

If you're a sports fan who thinks the world began with ESPN, you're in luck. Just do a Web search for:

```
sports
```

and pick the resources of your choice. Here you'll find lists of links to sports sites where you can read about sports scores, sports figures, and the latest in sports news. If you're more selective in your tastes, however, you can just search out specific sports or teams, as in:

```
football
```

or:

```
Big 12 Basketball
```

On the other hand, you might prefer actively participating to just watching and reading. Try a search for:

```
sports equipment
```

to find dealers and individuals on the Web, or try a Usenet search for something like:

```
golf bag "for sale"
```

standards

If your company is trying to become standards-compliant—which often means being ISO 9000 certified—you'll probably need to bone up on what it all means, as well as trying to figure out how to do it in the most convenient, least painful way. Try a Web search on:

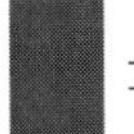

```
ISO 9000
```

If you want to hear the trials and tribulations of people who have already trod the path you're planning, head over to Usenet and search on:

```
ISO 9000 -newsgroups:jobs
```

The *-newsgroups:jobs* stipulation eliminates the thousands of jobs being offered to the people familiar with ISO 9000.

substance abuse

Having just had a near collision with a trailer truck on the highway, you are concerned about the issue of substance abuse and truck drivers (he *must* have been on something). You'd like to check to see if federal law requires any kind of drug testing of interstate truck drivers. You do an Advanced Search of the Web for:

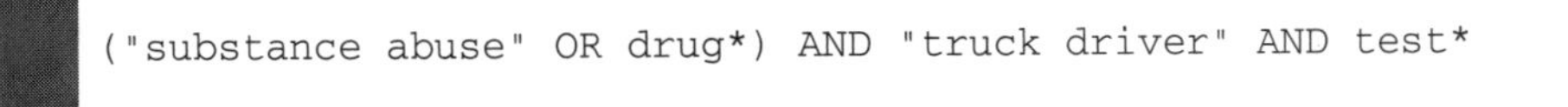

```
("substance abuse" OR drug*) AND "truck driver" AND test*
```

and in the Ranking field you enter:

```
law interstate federal
```

You get about 200 hits, and right at the top of the list is an article about qualifications necessary to be an interstate truck driver.

If you're looking for information about help for substance abuse, check out the section on "support groups" in this Reference for search ideas.

sugar packets

What in the world would you do with sugar packets? Well, if you search for:

```
collecting sugar
```

you'll find out that keeping a sugar packet collection is an interesting hobby. The question remains, though, should you save the packet *with* the sugar, or without?!

Or, if you're just looking for sugar trivia, try:

```
sugar wisdom
```

to find quips and quotes that appear on individual sugar packets and boxes. No kidding!

support groups

Many or most support groups that anyone would ever need are represented on the Internet. Even information for the most obscure support groups are found in Usenet. For example, enter:

```
newsgroups:alt.recovery.clutter
```

if you just cannot keep up with the clutter you keep accumulating. With this search, you'll find links to information on clutter, on containing clutter, on containing your containers, and even on getting rid of your clutter.

Or if your clutter is under control, try searching more globally for:

```
support groups
```

or for:

```
newsgroups:recovery
```

Certainly you can search on the Web for more support group information, like this:

```
"support groups"
```

but it tends to be somewhat less personal than the information on Usenet.

T

tests

If you're approaching the end of a course of education—high school or college—and considering moving on to something bigger and better (with presumably higher student loan payments), you'll probably have to take those awful standardized use-a-number-2-pencil-and-bring-two-extras type tests. Color in the circle for AltaVista and search for:

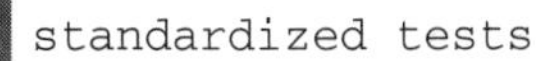

```
standardized tests
```

to get some information about the tests and some preparation programs. For more specific information, try:

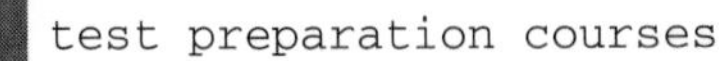

```
test preparation courses
```

or something specifically oriented to the test of your choice, such as:

```
LSAT preparation
```

or:

```
SAT preparation
```

In Search of Taxes

In reference to a question about taxes, Steve wrote, among other things:

> *Finally, if you have access to the World Wide Web, you might find some useful info. I did a search at AltaVista (http://altavista.digital.com) on the topic (using Advanced Search):*
>
> *501(c)3 or 501(c)(3) or 501c3*
>
> *sorted by: tax*
>
> *… and found approximately 4000 entries. At first glance, it appears some of them are useful.*
>
> *The IRS has a Web site from which you can download its publications and forms, including publications related to 501(c)3 status. The URL is*
>
> *http://www.irs.ustreas.gov/prod/forms_pubs/index.html*

—Steve Freedkin

trademarks

If you're going into business, you might need to check to make sure that your snazzy new product name doesn't infringe on someone else's trademark or registered trademark. These services probably cost some money, but likely less than a lawyer. Look for:

```
"trademark search"
```

and pick the service of your choice.

See also the "copyright issues" section of this Reference for more search ideas.

transportation (personal)

If you're tired of walking, hitching, or biking, or if the old Microbus just won't go another mile (or if you just think it's time for a change, reliability, or a new look in transportation), you can always use AltaVista to find out everything you need, *before* you have to deal with the car dealerships. Try a search for:

```
buying leasing new car
```

These links won't give you everything about the car itself, but they'll help you determine what do to—lease or buy. On the other hand, if you're looking for a used car, try a search for:

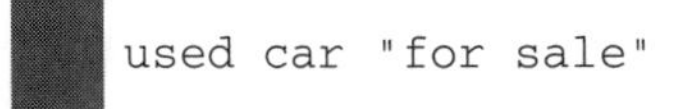

```
used car "for sale"
```

When you've nearly made your decision and you're almost ready to plunk down your money, take a quick time-out and search for the car by model and make on Usenet and look for any reports of problems or repair issues. For example, search for:

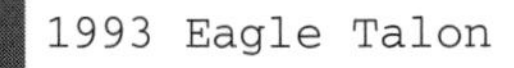

```
1993 Eagle Talon
```

and just see what's out there. You might find the car of your dreams without ever leaving your house, or save yourself from purchasing a lemon! Or, if the car business doesn't appeal at all, try:

```
hitchhiking tips
```

or:

```
bus schedules
```

travel

So you say you're vacationing in the Cayman Islands? Lucky you! Let AltaVista Search help with your travel plans. For starters, try:

```
"Cayman Islands"
```

This search gives you about 26,000 links to information about the islands as well as tourist attractions. Have you made your hotel reservations yet? Better try a search for:

```
"Cayman Islands" "Holiday Inn"
```

to head directly to the list of Holiday Inns on the islands. You still get thousands of matches with both those terms, many of which can help you plan your visit in some detail before you get there—for instance, comparing a wide variety of scuba diving packages. Then, of course, you need to know what to bring. Check out:

```
"Cayman Islands" weather
```

to see the four-day forecast and also tables of historical information on weather extremes. Better still, search for:

```
+"Cayman Islands" +"rainy season"
```

and you'll discover that the rainy season—with frequent, heavy tropical storms—lasts from May to October. (Whenever you're planning a tropical holiday and counting on lots of sun for outdoor activities, it's a very good idea to do a search for "rainy season" in that locale.)

You can even do newsgroup searches to get candid reactions from vacationers regarding the delights they found and the problems they encountered. For instance:

```
+"Cayman Islands" +problem* vacation
```

gives the paranoid a long list of problems to try to avoid—particularly related to scuba diving. Also, take a look at the "weather" section of this Reference for more search ideas before leaving for your vacation spot.

In Search of Tools

In reference to a question about finding Morton Machinery on the Web, Jim wrote:

> *Fundamental Web-browsing 101: Find a good search service—my favorite is Digital's AltaVista: http://altavista.digital.com/—and enter "Morton Machinery" as the query. The very first answer is Morton's Web page, http://www.mortonmachinery.com.*
>
> —Jim Kirkpatrick

trivia

If you haven't found enough interesting or bizarre stuff on the Internet yet, just look to the Web for collections of trivia contests with a search for:

```
trivia contest
```

The "Brady Bunch" contest should be a snap for the kids of the '70s, but weather trivia, along with most of the other contests, is fair game for a mixed audience.

If you're specifically interested in contests, check out the search ideas under the "contest" section of this Reference. Or if you have specific trivial questions, of course, just ask AltaVista Search—someone out there has probably already published the answer!

U

UFOs

Need to find out if that blinding light over your house was really a UFO? Read what everyone else has to say about them with a Usenet search for:

```
UFO
```

or possibly for:

```
UFO proof
```

Additionally, a Web search for:

```
+UFO +picture
```

produces a few links with images representing UFOs. Decide for yourself!

ultralight aircraft

So, you've just gotta do it. Do a search for:

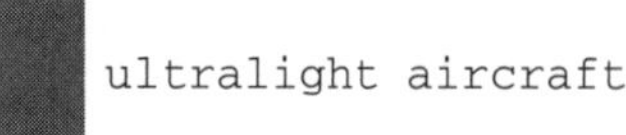

```
ultralight aircraft
```

and find everything from aircraft for sale and aircraft wanted to a whole slew of links, pictures, and other information.

urban legends

If you're excited about forwarding that Neiman-Marcus cookie recipe to all your friends, or in a tizzy about the Good Times virus, check out the urban legend information on the Web and Usenet. Specifically, do a Web search for:

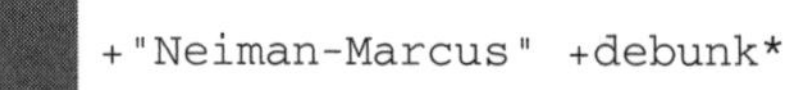

```
+"Neiman-Marcus" +debunk*
```

or:

```
+"Good Times" +debunk*
```

For a more comprehensive look at urban legends, check a Web search for:

```
"urban legends"
```

to get the real story behind dozens of hoaxes and bogus stories. Good discussions are also available on the Usenet side of things with the same search. Or you can try:

```
newsgroups:alt.folklore.urban
```

for more focused discussions.

V

vacation rentals

If you just have to get away from it all but don't really want to stay in a hotel, use either Usenet or the Web to hook up with a vacation rental house. As usual, add the specific location if you want, like this:

```
"vacation rental" +Bahamas
```

But it might be more fun just to browse (use *"vacation rental"* without the location). Of course, you'll probably need to take care of the pets and houseplants while you're gone, so try:

```
house pet +sitting
```

on both Web and Usenet. With these searches, you'll find loads of ideas for places to leave your pets as well as for services that offer house-sitting while you're gone.

volcanoes

Natural disasters can be fascinating—if you're far enough away, that is. A search for:

```
volcano
```

gives a whole collection of sites specifically on that subject. If you're looking for information on a specific volcano, just throw the name in as well, like this:

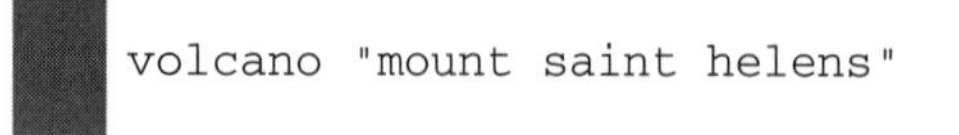

```
volcano "mount saint helens"
```

If you're so fascinated that you just have to seek out an academic program or at least a more scholarly approach to the phenomenon, try:

```
volcanology
```

This search provides links to volcanology programs at universities, information about the field, and even simulations.

W

wallpaper patterns

They're on your walls and on your computer—better find patterns you like! Not to worry. AltaVista can help you find wallpaper for both uses—easy and fast! If you're looking for wallpaper for your computer, try a Simple Search for:

```
wallpaper pattern
```

which gives you all kinds of links to interesting designs and patterns. Of course, if you're trying to find wallpaper for your bedroom, bathroom, or hallway, try this:

```
wallpaper pattern house redecorating -desktop -background
```

weather

Whether or not you like it, weather is here to stay. If you need to get the latest, search for *weather* plus your location, like this:

```
weather +Tulsa
```

This search gives you all kinds of choices about the weather information available for the location, ranging from (fairly) current conditions to the extended forecast. Additionally, you can get the big picture with:

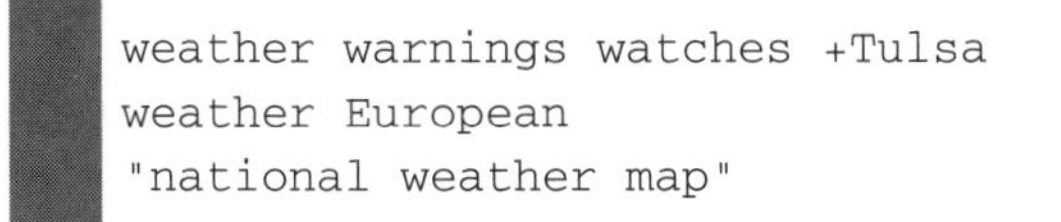

```
weather warnings watches +Tulsa
weather European
"national weather map"
```

or:

```
weather hurricane
```

Certainly, the weather sites cannot replace the accuracy and timeliness of your local forecasters. So if threatening weather comes along, check out what your local weather guy or gal has to say or turn on weather radio.

Web statistics

Big. How big? BIG! Maybe try something like:

```
measuring "world wide web" marketing +size
```

Or try a site called "Measuring the Web" (searchable by that name), which provides a remarkable picture of the interconnectedness of the Web, if you like that sort of thing.

women's issues

Women's issues and related topics comprise a good-sized chunk of the information available on the Internet. You can start by searching the Web for general information by entering queries like:

```
women's resources
```

or:

```
women's studies
```

With these searches you find everything from resource lists to support groups to public policy information. Similar searches on Usenet tend to be more focused but less productive.

Or you can narrow your search and find specific information. Suppose you are concerned about the issue of female circumcision in Africa and particularly in Ethiopia. Use Advanced Search of both the Web and Usenet, for example:

```
(clitorectomy OR "female circumcision") AND (Africa OR Ethiopia)
```

You get about 750 hits—most of them right on target. Or suppose you would like information regarding sexual harassment in Japan and in Japanese companies. You can search the Web for:

```
+"sexual harassment" +Japan*
```

and get about 2000 matches. The top or near-the-top items provide a good overview of the issue.

working at home

Tired of the old grind? Get up. Shower. Eat doughnuts. Go to work. Work. Go home. Sleep. Get up. Shower.... Well, AltaVista Search can't bring home the bacon for you, but it can help you find alternatives to your daily grind. If you want to work from home, try a Simple Web Search for:

```
SOHO
```

(yes, use all caps—SOHO stands for Small Office Home Office), or:

```
"alternative officing"
```

which give you links to home officing ideas. Other useful searches would include:

```
telecommuting
```

or possibly:

```
*"health insurance" "self-employ*"
```

world almanacs

Oh, come on. You need more information than you can get on the Web? Really? Oh, you say it needs to be portable. In that case, just do a Simple Search for:

```
"world almanac" "book of facts" ordering
```

Of course, a well-directed search for specific information usually suffices. Say you've just got to know the population of South Dakota. No problem! Try an Advanced Search for:

```
population and ("South Dakota" or SD)
```

and in the Ranking field, enter:

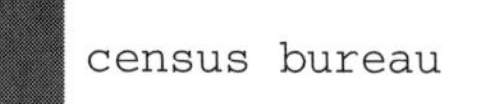

```
census bureau
```

X

Xena: Warrior Princess

Are you a fan of the syndicated TV series "Xena: Warrior Princess" but can't keep track of it because it keeps moving from station to station and time to time? A Simple Search of the Web gets you what you want quickly. Try:

```
+"Xena: Warrior Princess" +station*
```

This search gives the time slot (or at least the day) for this TV series and the stations it plays on.

xylography

Huh? You know, *xylography*—the art of making engravings on wood. If you're interested in some basic information about the art of xylography and its origins, try a Simple Search, like this:

```
xylography
```

Of course, if you can never remember how to spell it, you can always try:

```
wood engraving
```

Y

yachting

Okay, so you've had enough of running around with the riffraff. It's time to search for:

```
yachting
```

You can cruise through the links and choose the boat you want to charter. If you're really feeling flush, try:

```
yacht "for sale"
```

See also the "weather" section of this Reference for more search ideas related to yachting requirements.

Yellow Pages

Just as the Yellow Pages in your local phone book let you look before you buy, the Yellow Pages on the Internet let you do some looking and researching before you buy. Unfortunately, unless there happens to be a set of pages for your area, it's like picking out a set of Yellow Pages from Albany when you live in Boise—interesting, but pretty useless. If you're from Albany, try:

```
"yellow pages" +albany
```


Alternatively, if you already know what you're shopping for and haven't been getting enough catalogs in your daily junk mail deliveries, try:

```
catalog
```

You could also search for:

```
catalog -domain:edu
```

to exclude college catalogs—there's rarely anything to buy there.

Z

Zimbabwe

Suppose your company distributes used personal computers and you have heard that there might be good opportunities for sales in Africa, particularly in Zimbabwe. You'd like to know what the local market is like and would like to find a local company to partner with. You can start with a broad search, like:

```
+Zimbabwe electronics computer*
```

This search yields over 20,000 matches. The country may well be more developed than you imagined. Then you can focus the search a bit, like this:

```
+Zimbabwe +used +computer
```

This search gives you over 600 matches, but many of them deal with the question of how to connect to the Internet with a PC from Zimbabwe; and judging from the country codes, many of the Web sites are in South Africa (.za). You can focus the search even more if you're interested in information that resides directly on servers in Zimbabwe by entering a search with the country code (.zw), like this:

```
+host:zw +computer* +used
```

If you enter the country code, you don't need to enter the country name, in this case Zimbabwe. With this particular search, you get half a dozen matches—all pages at the same server, run by Samara Services. You note the email address of the Webmaster, and you also connect to the root address of the site (stripping away the directory and page portions of the address). From here not only can you get details about the company and its Internet services, but you can also find useful information about business in Zimbabwe.

zines

A *zine* (as in maga*zine*) is a publication, generally a periodical, about a certain topic. E-zines, as in email, are generally Web-based. Adding *+zine* to a search often brings the zines to the top, as in:

```
cars zine
music zine
```

or:

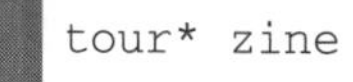

```
tour* zine
```

ZIP codes

What businesses and individuals located in a given ZIP code area are on the Internet? Try the Web and Advanced Search (to make it easy to enter all the possible variants of the state name):

```
(MA OR Mass OR Massachusetts) AND 02132 AND domain:.com
```

02132 is the ZIP code for West Roxbury, a residential section of Boston, where you would not expect to find many businesses using the Internet. But that search yields over 400 matches—people you might want to get in touch with to share experiences, or possible prospects for services that help companies do business on the Internet, or possible employers. Cool!

Zodiac

You can get your daily fix of astrology through a quick Web search for:

```
zodiac
```

and find your way to your horoscope as well as information about the planets in a more prosaic sense. Of course, an Advanced Web search for:

```
(planets or stars or constellation*) AND NOT (movie or film)
```

with

```
pictures
```

in the Ranking field gives you the best view of the stars this side of the *National Enquirer.*

zoos

Want to take your kids to the zoo but don't want to pile them in the car, wait in lines, or buy cotton candy? Let AltaVista Search take you on a safari! Check out all kinds of animals as well as actual zoos online by searching the Web for:

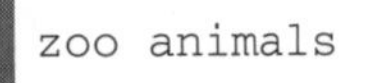

```
zoo animals
```

Your kids will see almost as many animals as they could in person. Or try something like:

```
zoo "St. Louis"
```

if you want to visit a specific zoo's Web site. Either way, you'll find an afternoon of fun, lots of information, and even help for the kids' school assignments. Enjoy!

zymurgy

If you're interested in winning bar bets, look up:

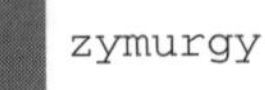

```
zymurgy
```

on the Web and find hundreds of matches about—drum roll, please—brewing beer. You'll find everything you'd want to know about it, from recipes to instructions to materials needed. If you're looking for a broader spectrum of information, you might expand the search to include less erudite terms with a search for:

```
zymurgy homebrew* brew* beer
```

Actually, taking zymurgy out of the search string might be helpful—that word is so unusual that it's going to bubble to the top of the matches nearly every time. And if you want to get down to the good stuff, try:

```
beer tasting
```

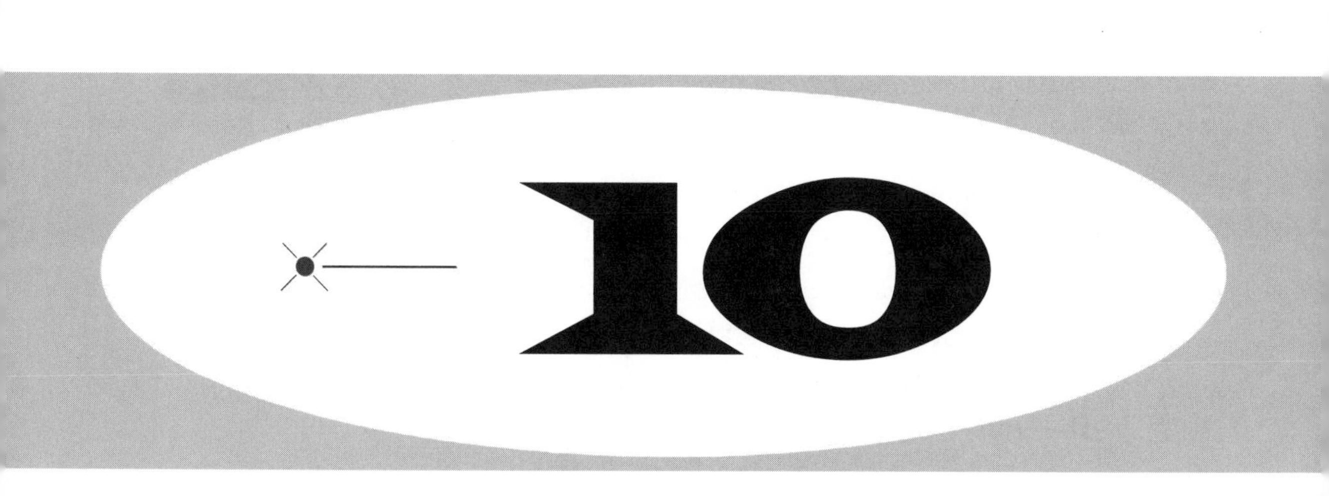

The AltaVista Story

The genesis of AltaVista—how it developed and quickly became one of the most widely used and respected Internet resources—is a tale of people, of technology, of innovation, and of many different perspectives on a large and wildly successful project. More than anything else, though, it is a story of synergy. Raging success like AltaVista can't just happen, or can it?

SO, WHERE DID ALTAVISTA COME FROM?

The story of AltaVista extends far beyond the obvious aspects of software and computers and networks—it ranges into areas such as Digital Equipment Corporation's commitment to research and development and Digital's often altruistic involvement with Internet technology. The serendipitous interaction among the various researchers and their diverse projects combined with Digital's technological advances led to AltaVista Search's development and ultimate success. In turn, the evolution of AltaVista created an entire Internet software business that markets and sells not only products and services based on the search technology, but also other Internet software applications.

In brief, the synergy among different aspects of Digital Equipment Corporation explains how AltaVista grew from a mere proving ground for incredibly fast computers to become the solution to the world's biggest information retrieval problem—the Internet. Then, with this synergy, AltaVista developed into the preeminent search service on the Internet, offering innovate ways to search, innovative features like on-the-fly translations of foreign language Web pages, and an ever-increasing size and scope.

AltaVista was not designed to imitate anything that had been done before. While it ultimately provides yet another example of how Digital leverages its expertise in the labs into real-world solutions to real problems, it was not designed with any specific business model in mind. To the researchers who developed AltaVista, the main purpose was to do what many considered impossible. The researchers needed to explore their ideas and push the limits of performance as far as possible and see where that would take them.

The AltaVista story provides the background and sets the stage for these interactions and describes the technology, the network, the programs, the people, and the project that led to the product now known as AltaVista. The story continues with the inside view of how such a massive service can run so reliably and so quickly, 24 hours a day, 365 days a year. The story concludes—but really just begins—with the next generation of AltaVista and the exciting possibilities that have evolved with it.

SETTING THE STAGE

AltaVista resulted from Digital Equipment Corporation's investment in research, emphasis on the Internet, and business attitude toward research and applied technology. The investments and commitments that eventually yielded the program grew from the development of the Palo Alto research labs, the ongoing emphasis on Internet-related technology, and from Digital's approach, which emphasized effective and quick use of innovative developments.

DIGITAL ON RESEARCH

Over the past twenty years, Digital's research endeavors have evolved into a much more structured and organized approach than was originally practiced. In the 1970s and 1980s, when industrial research first became a prominent part of large corporations, the Digital model fostered creativity in the form of pure research, without particular regard for business needs. The approach was to "gather very bright people, give them the best possible environment, then sit back and wait for magic to happen," notes Bob Supnik, vice president in charge of research at Digital.

Sam Fuller, the vice president who built the company's research organization, notes, "Back then, Digital was at the stage in its growth where it needed applied research capabilities second to none in the industry. We invested to develop our capabilities in important applied research areas to make sure we were at the forefront of computing."

As the information technology industry matured, Digital's research program moved from a model of undirected industrial research toward research based on market and business needs. The transition attempted to preserve innovation while maintaining a focus on the marketplace and the corporation.

Digital now focuses—like most other corporate R&D shops—on applied research. Rather than just retreating into labs to conduct research for its own sake, researchers focus their efforts on applying innovative concepts to solving engineering problems. According to Bill Strecker, Vice President and Chief Technical Officer in charge of corporate strategy and technology, "Research efforts must now have a practical realization to help the company develop, grow, and compete. Practical research, of course, requires that the research efforts actually result in a tangible product, so the researchers also build, use, and evaluate experimental prototypes."

This process is precisely what led to the development of AltaVista. The use of a research and development approach, sheltering the developers from the daily scrutiny of business units, made it possible for the AltaVista team to complete a prototype very quickly. It was less than six months from the time the researchers started to the time that AltaVista went live on the Internet and started serving millions of users. The first research investigations into very large memory indexes quickly led to investigations on retrieving and indexing immense amounts of information from the Web and finally led to the initial public unveiling of AltaVista in December 1995.

The resulting success heralded the development of additional features and improvements to the AltaVista software technology and services, which just goes to show that emphasizing research and development—even in unconventional ways—can have an enormous impact on a company or a product.

The very short time-to-market that the sheltered research environment made possible was essential to the success of AltaVista because the Internet changes so quickly. If it took an extra year to develop useful capabilities, by the time they were ready, Internet needs certainly would have changed. As it is, search services in general and AltaVista in particular today provide such an essential service that Webmasters go to incredible lengths to make sure their sites can be found easily and indexed well.

PIONEERING THE INTERNET

Researchers at Digital's Palo Alto Lab have been involved with the Internet from its earliest days. They frequently work in partnership with other Internet pioneers and have a tradition of doing whatever they can to help the Internet thrive. By way of example, for over ten years the labs have maintained one of the world's largest archives of free public software, available for anyone to download over the Internet at any time. Additionally, the Palo Alto labs serve as a major news backbone for the Internet. Digital's Palo Alto Internet Exchange (PAIX) provides one of the major traffic interchange points on the Internet for Internet service providers (ISPs) as well as network service providers (NSPs).

In fact, Digital claims a lot of firsts on the Internet that—looking back—are quite impressive. Did you know that Digital

- was the first Fortune 500 company to establish a public Web server?
- established the first Internet presence for a city in America? (Palo Alto, California)

- was the first computer company with an ARPAnet (the original Internet) site?
- was the first organization anywhere in the world to have two geographically separate gateways? (in Massachusetts and California)

And there's more! Did you know that Digital

- was the world's first computer company to register an Internet domain?
- created the world's first corporate Internet mail gateway?
- created the world's first commercial Internet print server?

We aren't done yet! Did you know that Digital

- is a founding member of the World Wide Web Consortium?
- was one of the first companies to utilize public Usenet newsgroups for customer support?
- was the first computer company to offer online ordering over the Internet?
- even today has one of the most active gateways on the Internet?
- pioneered micro-payment technologies on the Internet through the Millicent project?

Okay. Enough bragging—you get the picture. Far from being a single flash of brilliance, AltaVista is merely the latest in a long line of Digital's successful Internet projects as well as the latest in a number of services provided to the Internet community.

One of the predecessors to AltaVista was a marketing endeavor initiated in October 1993. As soon as Web browsers became available for PCs, the Palo Alto researchers made Digital's marketing information available on the World Wide Web. Thus Digital became one of the first Fortune 500 companies to actively use the Web for marketing, clearly demonstrating the business potential of the increasingly Web-centric Internet, while benefiting the corporation at the same time.

A series of "Internet showcase" projects soon followed this pioneer use of the Internet. These practical experiments used the World Wide Web to show the potential of that new medium, to assess what technology it would take to be effective, and to evaluate the business implications of the technology.

Early in 1994, the Palo Alto research team made theirs the first city on the Web. Later that year, the researchers took on the enormous task of providing live coverage

of the California state election on the Internet. In addition to providing the basic won/lost tallies, the team generated detailed graphics of results in all parts of the state on the fly—as quickly as the information was available in the computers. Maps and graphs of the results were also available through the site. That project clearly demonstrated the importance of the Internet for government and for elections in particular, and also pushed the limits of available technology in serving hundreds of thousands of users in a concentrated period of time, delivering not just text but complex graphical material to all those users, and doing so with excellent response, even at times of peak usage.

These and other projects were not sales efforts by Digital; rather, they were attempts to gain direct practical experience and to break new ground on the Web, and, not incidentally, to move the Internet a giant step ahead toward being a practical medium for marketing, sales, government, and a wide variety of business endeavors.

Researchers like Brian Reid and other Internet pioneers at Digital saw that the Internet was the future of computing and communication, that the Internet would transform how people and companies work and interact. They thought that in the long run, in ways that might not be immediately apparent, doing what's best for the Internet would be a sensible approach from a business perspective as well. They were right.

Building a Showcase

The AltaVista Public Service actually began in the spring of 1995. Louis Monier, a researcher from Digital's Western Research Lab, was talking over lunch with Joella Paquette, a marketing specialist in Digital's Internet Business Group. Paul Flaherty, from Digital's Network Systems Lab, joined them—and the rest, as they say, is history.

Paul Flaherty knew from personal experience that finding information on the Web was becoming increasingly difficult. It seemed that the Internet's usefulness would be soured by sheer information overload and complexity. Paul had an ambitious idea about how to demonstrate Digital's newest product, the Alpha 8400 (TurboLaser)—a computer that in tests had shown it could run database software a hundred times faster than the competition. "That was spectacular," he recalls. "You don't normally see that kind of leap in technology—two orders of magnitude in performance."

It dawned on Paul that a database could be used for Internet searching and that a machine a hundred times faster than anything else at database applications could do a much better job of searching than anyone had ever done. Paul wanted to build a database of the Web and provide access to it as a free service to everyone on the Internet.

Keying off this idea, Louis Monier envisioned a *full text* search of the Web, with every word from every page at every site available to searchers. According to Joella, "As the team explored this idea, it fueled our enthusiasm the more and more we talked. It sounded like a natural winner—likely to attract millions of users per day." As so often happens in a research environment, one great idea leads to more great ideas, and so came the notion of making the whole Web searchable.

Although this serendipitous meeting over lunch (at Left at Albuquerque, in downtown Palo Alto, by the way) sounds far too contrived to be plausible, it's most assuredly true. The unique atmosphere around the Digital labs—located in a variety of buildings, some historic, clustered within easy walking distance of each other around downtown Palo Alto, California—offers an ideal opportunity for Digital researchers to meet over coffee or lunch and share ideas or brainstorm.

Scooter

The software needed to search and automatically find and bring back information about Web pages—fast enough for the information to be current—hadn't been developed when AltaVista was conceived. This software would have to operate at unprecedented speeds of up to a hundred times faster than anything currently available and would have to retrieve millions of pages per day.

Programs that automatically visit Web sites and gather information are variously known as "spiders," "crawlers," "robots," or "agents," because they crawl around the Web looking for data. A program intended to visit not just a few select sites, but rather every page on the entire public Web—tens of millions of pages, and eventually hundreds of millions of pages—would have to operate at extraordinary speed to bring back pages fast enough for the information to be timely.

To appreciate the complexity of the problem, consider that a spider functions very much like a browser. Imagine setting out with a browser, clicking to a site, jotting down what you find, and checking to determine if you had seen it before. Then you click on another link and another link, each time checking to see if it is new. It would take time to get to each page and more time to get links to sites you hadn't been to before. And if you built a program that went through those same operations, nonstop, 24 hours a day, it might only get about a page a minute, or fewer than 1500 pages a day. At that rate, it would take such a program more than a human lifetime to look at 30 million pages just once—let alone the 100 million pages in the current AltaVista index.

Comments from the Inside: Philip Steffora

"I inherited a growing baby with a voracious appetite for TurboLasers," remarked Philip Steffora, AltaVista's Operations Manager, in September 1996. "When I joined AltaVista in April 1996, we were rolling the third and allegedly 'final' TurboLaser in the door. Five months later, we received delivery of four additional TurboLasers, bringing our total to eleven.

AltaVista has provided remarkable operational challenges for my team. We have now automated much of the monitoring and troubleshooting, leaving our late nights available for continued expansion and hardware installation. Not a week goes by that we don't add at least one new server or major piece of hardware. Our goals for the future include continuing to increase performance, reliability, and redundancy, while working to satisfy the ravenous hunger for more hardware."

Note *Now, after years of growth, AltaVista boasts 27 AlphaServer 8400 5/440 servers.*

So Louis Monier, leading the AltaVista project with the help of a variety of experienced Internet, software, networking, and hardware experts inside Digital's research labs, developed Scooter, the super-spider. Scooter was created from scratch with the sole intention of making AltaVista Public Service a reality. Louis not only produced Scooter in a remarkably short time but also simultaneously developed the Web front-end for Scooter and the indexing software that millions of people now know as AltaVista Search.

How Can It Work So Fast?

AltaVista produces results remarkably quickly—it's actually probably faster to find something on the Internet through AltaVista than it is to find something right on your personal computer through more conventional means. Part of the reason for the difference is, of course, the AltaVista software itself. However, the hardware that runs AltaVista Public Search—the query interface, Scooter, and the indexing software—is pretty impressive as well.

The Query Interface, what you see at **http://altavista.digital.com/** and probably think of as *being* AltaVista Search, runs on five identical Alphastation 500/500s, each with 1GB RAM and 13GB of hard disk space. Each of these systems runs a custom multithreaded Web server that accepts your queries and sends the queries along to the index servers. These systems, relatively small by AltaVista standards, forward millions of hits per day to the AltaVista index servers, with roughly 90 percent of the queries going to the Web and the remaining 10 percent to newsgroups.

The real basis for AltaVista Search's superior performance lies with the four sets of four AlphaServer 8400 5/440s systems, each with ten processors, 8GB of RAM, and 450GB of hard disk in a RAID array. (RAID systems are sets of hard disks that work together and ensure that if a hard disk fails, no information is lost.) Each set of four servers (quad) holds a complete copy of the Web index with 1/4 of the index on each of the four machines (currently 75GB per machine) and provides a response time of less than a tenth of a second through the AltaVista system. In addition to the security provided by having four complete copies of the index running at all times, a complete additional set of four identical machines is available offline and can be pressed into service if necessary.

Scooter runs on an AlphaServer 4100 with 1GB RAM and a 120GB RAID array. All Scooter has to do is roam the Web, retrieve information,

and send it to the indexing system, which compiles the index. Vista, an AlphaStation 8400 5/440 with 10 processors, 8GB of RAM, and 450GB of hard disk space in a RAID array, actually builds the index, and then copies it to the index servers.

The news server runs on an AlphaServer 4100 system, with 1GB of RAM and 75GB of RAID hard disk. This server keeps the news articles spool for the News Indexer and serves the articles to users who want to read news with their standard Web browser. The News Indexer runs on an identical machine. The old news server and indexer—AlphaServer 600 systems with 256MB of RAM and 75GB of hard disk space—act as hot spares for the current server and indexer. These machines maintain an up-to-date index of the news spool, which includes handling the constant turnover of thousands of news articles as some expire and are deleted and others arrive in the system.

A Database or an Index?

Even with Scooter's help in retrieving all of the available Web pages and pouring them into a database, getting results back out again would be a challenge. Developing a real database would require a very large computer application that could record key bits of information about every Web page on the Internet. With such an enormous amount of data to search through—that is, data about everything on the Web—if people made simple one-word queries using significant words, they'd get many thousands of responses, which would not necessarily be useful. In addition, there was no way that a traditionally structured database approach could provide the fast response users would require.

Fortunately, software for an alternative approach—high-speed full-text indexing—had already been developed by another researcher in Palo Alto. The indexing software seemed a good match—the only challenge was that the index would require that the *entire Internet*, for all intents and purposes, be reproduced on computers at the Palo Alto labs. Digital's indexing software started out humbly as a way to organize email, and then, because it worked so well, evolved into an indexing system for the newsgroups, but the newsgroups still weren't nearly the same scope or size as the entire Internet.

The experience with these earlier projects helped the researchers understand, despite conventional wisdom to the contrary, that indexing the entire Web might be feasible. The indexing software was overhauled at about the time that the AltaVista Public Service project was taking off, and the huge set of data that Scooter was retrieving from the Web proved to be the perfect test of the new software.

A rough estimate at the time indicated that all the text on the entire Web probably amounted to less than one terabyte (a thousand gigabytes). In other words, the Web wasn't quite monstrous beyond imagination—every single word could be indexed, which would make the new search engine far more powerful and useful than anything else in existence. Additional work was needed so the indexer could handle the tens of millions of pages of text on the Web fast enough, but the target was within range. Although the rest of the world continued to think that a true complete index of every word on the Web was, for practical purposes, impossible, Louis and the team knew that, despite the difficulty, it could be done with off-the-shelf computer hardware and their own custom indexing software.

So within a few weeks after beginning work on this project, the AltaVista team decided that the correct approach was not to make a classification system using search terms in a database, but rather to build a full-text index of the entire Internet. The indexer would be far faster and far more powerful than a database, able to search for complete phrases and sentences and to link words in complex queries at no loss of speed. (Leaping tall buildings is currently in beta test.) And the AltaVista team would manage the anarchic, unstructured information of the Internet by using a tool that had been designed for dealing with large masses of unstructured information—a perfect match.

Developments in the past couple of years have proven, once again, that once a technical problem has been surmounted once, it's far easier for others to do the same thing. Now other companies offer searches of the whole Internet—full-text searches, in some cases. However, the advances the AltaVista team has made in search technology continue to force others to attempt to catch up with the lead set by AltaVista.

How Does the Indexer Work?

First, the indexing software takes the text of a document and examines every word in it to create an index organized by word. It saves each instance of each word along with the URL of the page on which it appears, and information about its location in that document. This level of detail is necessary in order to do phrase searches, which depend on knowing the exact order of all the words within a document. That is, AltaVista must know that a document includes "to be or not to be" in that order, as opposed to merely remembering that a document has the words "to," "be," "or," and "not."

AltaVista indexes all instances of all words, regardless of capitalization, as lowercase, and additionally indexes all words with capitalization again, exactly as their typography indicates. That allows users to do a general search or to narrow the search for unique capitalization, as in trademarks or other words with unusual capitalization, like *AltaVista*.

Similarly, it indexes under the English letter equivalent to all instances of words with accented letters from non-English languages that use Latin characters; so, Èlèphant is also indexed as elephant. Once again, at the cost of enlarging the index, this approach gives users considerable flexibility in focusing their searches.

Additionally, no order is imposed on this enormous body of information to make it accessible. AltaVista simply indexes words. It takes the unstructured content of the Internet and, without adding some arbitrary or human-designed structure or categorization, makes it easy for users to find what they want.

Making It Happen

Once the decision was made to use the existing Digital indexer, Louis took charge of the AltaVista project. Louis, naturally charismatic and personable, was (and is to this day) very excited, bordering on messianic, about AltaVista Search. His contagious enthusiasm got lots of people in the labs involved, through soliciting suggestions and criticisms and leaving the door open for others to add their opinions and comments. Louis notes that he "started from scratch," with no knowledge of networks. The top people in the field, however, were readily available right there at the labs, and he could drop in on them at any time and ask about any aspect of the

Internet or the Web or HTML (the text markup language used to create Web pages). He could get a "brain dump" from the experts and in a few hours know what he needed to know, or ask a targeted question and get the answer right away.

Researchers in the System Research Center as well as other Digital research facilities, including the Western Research Lab and Network Systems Lab, have also made substantive contributions to many aspects of AltaVista.

Glenn Trewitt and Stephen Stuart, researchers from the Network Systems Lab, were part of the core team that provided the networking expertise, designing the networks and putting together the hardware needed to run AltaVista Service. With their experience in network design and topology mapping, they set out to ensure that AltaVista would have the hardware and network connectivity needed for a successful debut. (Stephen continues to administer Digital's showcase Internet gateway in Palo Alto.) Before the AltaVista launch, the connectivity was somewhat over ten million bits per second—probably not enough to handle the load when AltaVista Search's customers showed up on launch day. With just a few months' warning, Stephen, Glenn, and others worked to upgrade Digital's connection to the Internet to over 135 million bits per second by the day of the launch. And this has grown many times since AltaVista appeared on the scene.

Whenever Louis needed equipment, which happened frequently, he could go to the system administrator, Annie Warren, and ask for a spare machine. And repeatedly she was able to come up with what he needed. Research facilities don't budget exactly for what people expect they'll need. The unexpected happens all the time, so spares are set aside for projects such as these—even when the spares are million-dollar computers.

"Remember," notes Annie, "we did this while running at full speed. We've never had the luxury of saying, 'This is what we're going to do. This is what we'll need. Now let's order the hardware for it.' We've always been doing 100 or 200 percent more than we ever thought we could with the hardware and trying to retrofit everything so we can eke out that tiny little bit more. We've been doing the equivalent of fine-tuning a race car engine while it is in the race."

Piece by piece, Annie gave Louis all her hoarded just-in-case equipment. Additionally, many of the researchers provided spare systems or memory cards to give Louis the hardware necessary to make AltaVista work.

Louis emphasizes that the Digital labs were probably the only places in the world where this work could get done so quickly. No startup company could afford the research staff. No university could have afforded the investment in equipment. And other large companies have not consistently invested in research the way Digital has, and thus could not reap the benefits of research investment.

Architecture of the Future

AltaVista is the problem Alpha was designed to solve. It demonstrates the capabilities that were built into the architecture back in 1988-89.

—Bob Supnik, Vice President, Digital Research Laboratories

Alpha computers—the computers on which AltaVista is built—use something called 64-bit architecture, which is a key factor in their performance. In contrast, most common personal computers today use 32-bit architecture, and often use software designed for 16-bit systems.

The number of bits designates how large a "word" the computer can handle at once. If everything else is equal, bigger words yield better performance. If the computer does a specific number of calculations per second but can consider more information per calculation, the performance improves. (It's exactly the same reason that firefighters use great big hoses instead of garden hoses—they can get more water through a bigger pipe.)

Alpha systems use 64-bit words—a very large pipe—which increases performance in many AltaVista operations, particularly calculating and comparing the information necessary to perform phrase and proximity searches. Additionally, using 64-bit words dramatically improves the way Alpha computers store and retrieve information, resulting in proportionate improvements in performance. If a typical query takes 2/10 of a second instead of 20 seconds, a search engine is transformed from a curiosity to a reliable, everyday tool.

Ten years ago, 32 bits were quite sufficient, even for high-end computers, because the amounts of information that computers were called upon to manage were far less than they are today. To handle tasks involving many gigabytes and even terabytes (a thousand times bigger than a gigabyte) of information requires the extra capacity and power of a 64-bit machine. And as an even greater proportion of the world's information is stored in digital form and made accessible over the Internet, computer systems will need to address, manage, and process petabytes of information (a thousand times bigger than a terabyte)—still well within the capabilities of the Alpha's 64-bit architecture.

Memories

AltaVista shows the practical value of machines that have more than 4 gigabytes of physical main memory.

—Dick Sites, Digital Equipment Corporation Systems Research Lab

In addition to 64-bit architecture, AltaVista's Alpha computer systems also use large amounts of memory, which helps improve the speed of your searches. Computers store information in two forms: the main memory (RAM), which is extremely fast and readily available for processing; and disk storage, which is about ten thousand times slower but is much less expensive and has a much larger capacity. Computers that hold more information in RAM perform better than those that hold less information, if everything else is equal.

High performance, which is essential given the incredible demands placed on AltaVista Search, requires holding as much information as possible in main memory—preferably the information that is most frequently used, so the system doesn't have to waste time fetching new items from disk. To get a sense of the relative size of the RAM in AltaVista Search, consider that a typical personal computer today comes with 32MB of RAM, and the AltaVista Index Servers each have 8GB of RAM—about 250 times more.

With 8GB of memory each, the AltaVista Index Servers can hold in main memory large chunks of the index—including document descriptors such as title, URL, and abstract. That means far fewer disk accesses, and thus machines that can provide fast service to millions more people per day than would be possible with less RAM.

BUILDING THE PROTOTYPE

The seed of the idea that became AltaVista was planted in April 1995. By June, Louis was writing the code for the Web crawler—Scooter. Meanwhile, work was proceeding on the indexer, fine-tuning that software in anticipation of an enormous new set of data that would soon be coming, and scaling up the performance by a factor of ten to accommodate the amount of information Scooter would retrieve.

The challenge in building Scooter was figuring out how to get lots of documents very quickly. Scooter needed to be able to fetch the text of the entire Web quickly enough so information in the index would be reasonably current. The solution was using a "multithreaded" program—the equivalent of having a thousand people browse the Web simultaneously. But that led to complexities in coordinating the activities of a thousand threads, all launched from one machine and one program, and taking all necessary steps to maximize performance.

Because the Internet is so unstructured and so many people who create Web pages don't follow all the standards for documents, nobody could precisely predict performance without live trials. There was no way to imagine the myriad of complexities that Scooter would encounter. Scooter had to be designed and redesigned, time and again, throughout the summer, with input and comments from Jeremy Dion and networking experts from Digital labs.

Scooter's first few limited trials occurred in June 1995. Over the Fourth of July weekend, Louis released Scooter to begin its first big crawl. Louis wanted to be sure that his creation didn't bother too many people before he'd had a chance to perfect it, and that holiday weekend is typically a low-traffic time on the Internet because so much of the Internet is still based in the United States. Louis started with a few URLs and let Scooter loose. At that point, Louis was running with a limited number of threads and constantly experimenting with the code to optimize performance.

"With a spider, if you have a bug, no matter how small, you are going to hear about it a lot because your spider will affect a lot of sites," observes Louis. "When you use multiple threads, that's like launching many different spiders at the same time, and your code is soon very visible to the whole planet. That reality forced on me a sense of responsibility that I've never had before." In order not to harass or trouble people, Louis built special safeguards into Scooter as well as obeying all commonly accepted conventions for spider behavior, such as the Robot Exclusion Standard. In compliance with the Robot Exclusion Standard, Scooter checks a special file at each Web site before visiting any of its pages. This file may contain a list of directories and pages that the site's Webmaster does not want robot programs to visit. Scooter does not fetch any of those. (See Chapter 8 for more information about controlling what Scooter visits.)

By the end of August 1995, Louis was ready for a big, full-scale crawl, and this time Scooter brought back about ten million pages. This experiment and the previous ones confirmed the supposition that the Web was not quite the monster that others thought it was. Extrapolations from what Scooter had found indicated that the Web probably consisted of over 80,000 servers and over 30 million indexable documents at that time, with the average document containing just five kilobytes (the equivalent of a couple of pages of printed text). Later, when the system went into production mode and Scooter was allowed to search farther, these estimates were refined upward to over 200,000 servers and about 50 million Web pages—but that was still in the same order of magnitude. In other words, as they had suspected, a full-text index of the entire Web was small enough to be manageable. Suddenly, the future of the Internet and how people would use it looked very different.

Piping Information into the Internet

We don't so much connect to the Internet, as we have it in our living room.

—Stephen Stuart, Researcher, Digital Equipment Corporation

For the AltaVista Public Search Service to maintain consistently high levels of performance and serve millions of users per day requires not just powerful computer hardware, but also extremely fast Internet access. Fortunately, even before AltaVista Search's appearance on the scene, Digital already had an enormous "pipe" coming from the Internet into the Network Systems Laboratory in Palo Alto, California.

When you're talking about networks, the unit of measure you use is megabits (millions of bits) carried per second. An average company might lease a T1 line from a telecommunications company, which provides a "bandwidth" of about 1.5 megabits per second, and in many cases that could be enough to meet the needs of an entire company. Very large companies and those that resell Internet service to other companies might lease one or more T3 lines, which provide about 45 megabits per second each. T3 is the standard for building a serious, continental-size network. The Network Systems Lab built Digital Equipment Corporation's Palo Alto Internet gateway with connections to Internet service providers that operate at more than twice T3 speed—100 megabits per second—using technology known as "switched FDDI."

THE PITCH!

On September 29, 1995, Louis made a presentation to Digital corporate management on the East Coast, pointing out that the search services then available on the Internet were either a table of contents based on manual classification or a book-type index based on selected significant words. He explained that AltaVista would explore the Web, find and fetch each page, read the contents of each page, find all the words, and add them to an index that would be accessible throughout the Internet.

One index would hold every word for the entire Web. Scooter could fetch pages at a rate more than a hundred times faster than anything then available. The full-text indexing software could index these pages as fast as Scooter could fetch them and provide a complete representation of all the text on the Web. It could very quickly handle queries that included phrases, complex combinations of terms, and unique structural elements of Web and newsgroup articles. And it would rank matches, so pages that were most likely to be useful would appear at the top of the list.

This project would let Internet users find anything on the Web. If searchers knew what information they wanted to find and if the words were anywhere on the Internet, the matches would be found.

Louis emphasized the value to Digital of creating the "ultimate" Web indexing site and described an internal pilot he was already rolling out for use by Digital employees. He was shooting for December 15, 1995, as a public launch date.

As a result of this presentation, the project became highly visible throughout the company. How would the world react to the AltaVista Public Service? How would they view it compared with other Internet search services? What should be done to ensure the success of AltaVista Search? What was the marketing plan? How could this service generate revenue? What about related services?

While there were many questions to finalize, lots of enthusiasm and support came from all over the company to help AltaVista Search's launch in the form of hardware, planning, and marketing and communications expertise. According to Jay Zager, then finance manager for Digital's Advanced Technology Group, "the competitive advantage was clear, as was the need to move quickly. After all, on the Internet, extraordinary change can take place in a matter of days, and startup companies are legendary for their rapid response." Digital could not afford to follow its ordinary business procedures in this case. They needed to position this project so it could move ahead with a minimum of hassle and delay.

In fact, the company had just announced its corporate strategy to the press and analysts, and the Internet was an important part of that strategy. But in the marketplace, even with all of its pioneering Internet history, Digital was still not perceived as a major Internet player. Sharon Henderson, then communications director for the Advanced Technology Group, recalls, "Digital had been in the Internet business for years, but I could call up any press person at a major publication and they couldn't tell me what Digital was doing on the Internet. The problem was that our servers and software were used in the back room. They weren't things that end users could touch and feel. Here was an opportunity to reach the broadest possible audience. My goal was to make sure that the message didn't just go to people who already understood the Internet, but also my grandmother or my father who could use it. In terms of visibility, this could be the biggest announcement that Digital had ever made."

So instead of quietly releasing the technology to the Internet user community, Sharon pushed for bringing in the big guns from corporate public relations and making a major announcement in the press. The decision was made to stick with the December 15 launch date.

The Internal Pilot

While preparing for that management presentation, Louis and the team were also getting the internal pilot up and running. They fed the ten million pages that Scooter had found to the indexer and then provided a search service for Digital employees as a live test of the capability. The pilot lasted for about two months and involved about 10,000 employees, who posted an average of about 10,000 total queries per day. Now, as internal testing proceeded, the developers had to deal not just with the diversity of the unstructured Web content, but also with the diversity of behavior and preferences of 10,000 individuals at Digital, many of whom were quite vocal and very helpful in making their preferences known.

This internal testing turned up a number of bugs with the user interface, most of which involved compatibility with a plethora of existing Web browsers. The development team struggled to understand why particular browsers repeatedly had trouble with certain features.

It was at this stage that the AltaVista development team decided to include both "Simple" and "Advanced" Search. Simple would be an intuitive search format allowing users to pose queries in plain language—any Western language—and immediately get useful results. Advanced, on the other hand, would be the native query language of the indexing software, using the terms of Boolean logic, such as *and* and *or*, which many technically oriented people are already familiar with and sometimes prefer.

"AltaVista" was the code name for the project. Now that AltaVista was going public, much discussion went into deciding if the name and graphics were correct. It is a little-known fact outside Digital that until two days prior to the launch of

AltaVista Public Service, the name and graphics were going to be completely different from the AltaVista that millions of Internet users know today. One of the early recommendations made by Ilene Lang, then executive-in-charge of the project, was to keep the AltaVista name.

Meanwhile, another internal pilot project at Digital demonstrated the value of an AltaVista Service for *intranets.* Intranets are corporate networks that use Internet networking technology and techniques to deliver information to and interact with employees and selected suppliers, customers, and partners in a secure, closed environment. These networks use hardware and software products known as *firewalls* and *tunnels* to block out intruders and facilitate confidential communications among trusted parties.

Digital's intranet is used by all of the company's 60,000 employees. It was originally assumed that there were about 300 internal Web servers—which seemed like an enormous number compared to the few external Web servers that the company used to deliver information to the public. Running an early intranet version of Scooter and AltaVista Search, the company discovered that it in fact had over 1100 internal servers and over 900,000 Web pages. AltaVista would not only make this information far more accessible to the people who needed it, but also show AltaVista Search's value as a diagnostic tool—helping companies better understand how Internet technology is used internally and helping them to make better use of their computing resources.

"The value of data appears to be greater the more of it you have," Brian Reid, head of Digital's Western Research Lab adds. "If you get twice as much data, it has more than twice the value. The increase in value is greater than linear simply because of the higher probability that you will find what you are looking for, and the greater likelihood that you will ask in the first place. A library that only has a hundred books is not worth going to. A library that has every known book is always worth going to. And the more information it has, the more valuable it is in proportion to the actual amount of information. So AltaVista Search, by enabling you to find things, increases the value of the information that you have on file."

Comments from the Inside: Annie Warren

"So much of what we do in research goes into high-level design that's invisible to most people," she reflects. "Now it's fun to work on something that you and your friends and relatives can personally use and enjoy. I got my dad to buy a PC after he used AltaVista. I got another friend to actually sign up for an Internet account, someone who is not computer literate at all, but she does genealogical searches. I brought her here to the office and showed her; and all of a sudden, she needed a PC and an account with an Internet service provider. Now she spends all her days doing genealogical searches using AltaVista instead of making her annual pilgrimage to Utah to stand there at books looking through birth records. It's so much fun for me to be able to say to my friends who are not in the computer industry, 'Look, this is what we do, and you'll really like it.'"

THE LAUNCH!

On December 15, 1995, AltaVista opened to the public. The team working on AltaVista saw the launch as an experiment. If it was successful, there were enormous product and service possibilities based on the search technology.

The launch was *very* successful. Usage grew dramatically—from 300,000 hits on the first day to over 19 million a day nine months later and 80 million hits a day two years after launch. A more telling—and reliable—measure of AltaVista's popularity is the fact that 15 million users use the service every month. Early users returned again and again and quickly spread the word about its many possible uses. Thousands of articles appeared in newsgroups of all kinds telling of personal experiences and sharing the enthusiasm of discovery. Awards and stories from all

over the planet began to flood newspapers and magazines about AltaVista Search. Finally, anyone who had access to a Web browser had an easy way to find information on the Internet.

The Growth

AltaVista's innovation and development continues and, if possible, even surpasses the initial success of the service. Throughout the first two years of AltaVista's existence, growth was the key word to describe the changes—growth geographically, growth in scope, growth in size, growth in power, and of course, growth of popularity.

With the tremendous growth in popularity of the single AltaVista Public Service in Palo Alto, California, the natural course of action was to deliver more of a good thing by mirroring the AltaVista Public Service site in many geographically dispersed locations throughout the world. These localized sites for AltaVista provide users with several advantages, including much faster searching (because the query doesn't have to go all the way to Palo Alto to be processed) and local-language AltaVista interfaces (so people who do not speak English do not have to try to puzzle out what the AltaVista tips and help say). The AltaVista Network continues to develop with new AltaVista mirror sites including, at most recent count:

- AltaVista Canada at **http://www.altavista.ca/**
- AltaVista Asia at **http://altavista.skali.com.my/**
- AltaVista Australia at **http://www.altavista.yellowpages.com.au/**
- AltaVista Latin America at **http://www.altavista.magallanes.net/**
- AltaVista Northern Europe at **http://www.altavista.telia.com/**
- AltaVista Southern Europe at **http://www.altavista.magallanes.net/**
- AltaVista Palo Alto (the original) at **http://altavista.digital.com**

Although all of the mirror sites have substantially the same information as the main site in Palo Alto, there is a slight delay in getting the newest additions to the index out to all of the mirror sites. If you're looking for information that was likely just added to the index and you can't find it at a mirror site near you, then try the main site in Palo Alto.

An additional feature now provided by AltaVista Public Service lets visitors to a site search the Web using the AltaVista engine without actually having to link to the AltaVista home page. Yahoo, one of the Internet's first guides and most popular directories, is a good example: Yahoo users' searches are now powered by AltaVista software whenever they want to search the Web. For example, if you search for something that simply isn't in Yahoo, you'll immediately see the results of an AltaVista search for the same thing—so you don't come up empty-handed.

Where else is AltaVista at work? As a result of the visibility and usefulness that Internet technologies have brought to the public arena, many corporations are recognizing the value of adopting the same capabilities inside their operations on their intranets. Most larger companies have set up networks using the same technology as the Internet and set up standard Web browsers on computers to provide access to corporate information. As the number of corporate intranet Web pages grows ever more rapidly, businesses will find that spending time just pursuing information is not as productive as using that information.

Digital's researchers and business analysts decided that putting technology behind the Internet search engine to work for today's corporations, universities, and government agencies would be tremendously helpful. Other AltaVista products extend the public search engine technology to private business environments, slightly adapted for locating various types of useful information on corporate intranets, personal desktops, or any file server a desktop is connected to—in seconds. If you're struggling to find information on your intranet or personal computer and wish that it were as easy as finding information on the Internet, know that it is possible.

In fact, in July 1996, an early release of a desktop version of AltaVista was made available for free trial downloading over the Internet, and newer and updated versions (under different names) continue to be made available. Imagine being able to instantly find data buried in an old email message or to find a file you thought was lost—accidentally saved deep in an obscure directory. With this tool, you can preserve all your work without complicated filing and naming systems and directories, and without worrying about how you'll be able to find it later. This makes it possible to find information on your own computer as easily as you can over the Web, with the same AltaVista look and feel and the same commands. Now entering its third version, AltaVista Discovery (a new name to go with its new look) offers an innovative and easy way to search your computer.

KEEPING IT GOING

Although the obvious excitement in the AltaVista story revolves around new developments, new features, and new directions, one of the most often overlooked aspects is the technical achievement of *providing* the AltaVista search service, day in and day out, 24 hours a day, 7 days a week, 365 days a year. Phil Steffora has led

the AltaVista Operations team from the very beginning—this is the team that makes it possible for you to search AltaVista whenever you want, and always get prompt responses and reliable information.

What makes it possible to keep AltaVista so remarkably reliable? The AltaVista approach to system administration could—and probably should—be used as a textbook case of the best way to run mission critical systems. First, though, a little background. When AltaVista began, the Alpha servers were housed in a machine room in one of Digital's Palo Alto research labs. The overwhelming popularity of the service almost immediately required additional servers, and the growth continued unabated. Unfortunately, rather than being able to grow and develop the physical AltaVista server site as need dictated, the site was constrained by the facilities.

For the first two and a half years of AltaVista's existence, it continued to grow and expand throughout all available space in the lab in downtown Palo Alto. Each additional server that was added further taxed the limited space—essentially evicting the hardware used by other Digital researchers—and the cooling capacity of the building. Eventually, the machine room was as full as it could be, but additional servers were still needed. At times, million-dollar Alpha servers were placed in hallways and stairwells and anywhere else that spare room could be found.

Of course, limitations on space and physical capacity did not alleviate the demand to maintain the systems and keep the reliability as high as possible. As you might expect, people who needed information from AltaVista weren't interested in excuses or stalling—they expected and received search results promptly, always.

However, the processes needed to ensure this reliability were more than a little interesting. For example, the UPS (Uninterruptable Power Supply or battery backup system) for the lab was well beyond capacity and could maintain the AltaVista servers (and assorted other equipment in the labs) for merely 15 minutes. A power outage of 16 minutes or more would assuredly bring the systems down. In order to forestall this possibility, Phil and the operations crew wired a new outlet directly into the power supply coming into the lab building and rented a large generator, permanently mounted on a trailer and parked a few blocks away. When the power to the building went out and the backup power supply kicked in, all operations staffers were automatically notified by pager and the nearest person ran for the truck and got the trailer.

When the trailer-mounted generator was brought to the building, there was no established parking place, and the on-street parking in front of the building was usually occupied. So, the trailer was double-parked or even pulled directly up on the sidewalk, as needed. They plugged the generator directly into the building's electrical wiring, started the generator, and AltaVista kept right on serving queries.

Now, of course, the AltaVista servers have their own home—designed specifically for the unique needs and demands of such a large, high-traffic operation. In addition to the more or less obvious features of incredibly tight physical security, ample space for growth, and backup power supplies, the levels of redundancy and reliability surpass even the wildest dreams of most system administrators.

Clearly, the specific features can't be described, but the rule of thumb used at AltaVista is to have at least one extra copy of everything ready at all times and to run no computers beyond 50 percent of their capacity. This extends from the smallest components (using two disk drives whenever only one is technically needed) up to the entire server system. A full, current copy of the AltaVista index (and needed servers) resides at a completely separate location from the main site but works parallel to the site in real time—if some problems take the main site offline, the backup site can easily pick up all of the traffic and move along seamlessly.

Similarly, multiple servers, multiple RAID hard drives, multiple connections to the Internet, and multiple people monitoring everything ensure that AltaVista remains just as reliable as its customers require.

The site is so large that even statistically unlikely events happen regularly. For example, on the average, hard drives go hundreds of thousands of hours without failures. Of course, some go longer, and some don't make it quite that long. AltaVista loses one or two hard drives each week to failures, but the levels of redundancy in the system allow the operators to simply swap out the drives (without taking the system down) and keep right on chugging.

Finally, far from the days—only a short while ago—of driving the generator onto the sidewalk, AltaVista now has four separate levels of backup power, ensuring that power outages can never again cause that much stress for the operations team at AltaVista.

EXTENDING SUCCESS

Where does AltaVista go from here? Part of the answer to that question is a secret—watch the AltaVista site for more information—but the other part is continued innovation and partnership. Through strategic partnerships and resource sharing with technology leaders in other fields, AltaVista's site continues to ride the cutting edge of Internet technology. Some of the interesting innovations over the past couple of years include Refine AltaVista's Filtered Search, developed in partnership with NetShepherd, Inc. and the overwhelmingly popular Translation Service, using translation technology from Systran, Inc.

The Refine Feature

AltaVista's Refine feature, covered in Chapter 4 of this book, resulted from a collaborative effort between François Bourdoncle of Ecole des Mines de Paris and the AltaVista team. François, a friend of Louis Monier's from years ago, took a leave from his position in Paris and went to Palo Alto to work with Louis and the rest of the AltaVista team on a new concept in searching. This approach uses a sophisticated statistical analysis process to automatically categorize Web pages into groups, based on their content. Rather than relying on predefined categories, as Web directories and navigational aids often do, Refine constructs topics on the fly to meet the particular content found on the Web in response to the search. Refine provides searchers with a high-level view of the topics and categories a search generates to readily include or exclude entire groups of information and easily focus a search.

The Filtered Search Feature

AltaVista's Filtered Search (accessible at **http://family.netshepherd.com/**) uses independent site rankings to rank and categorize Web sites. Only categorized sites that are appropriate for "family" viewing, in the collective opinions of the NetShepherd rating team, will appear in any matches from a Filtered Search. In addition to the obvious safety of being fairly sure that you won't have any surprises when searching, you can also take advantage of the rating system to choose only sites with the quality level, on a scale of one to five stars, that you want. Filtered Search takes advantage of the full Web searching and indexing capabilities that AltaVista offers, plus the PICS (Platform Independent Content Selection) standard-complaint NetShepherd database.

Language Recognition

AltaVista's language recognition features, developed with the help of Franz Guenthner, a linguist from Munich, Germany, automatically analyze all Web pages as they're indexed and label them by language, lets searchers easily select just information in specific languages, rather than getting flooded with information on a specific topic written in a variety of different languages.

The Translation Service

AltaVista's Translation Service provides on-the-fly translations of Web pages between any of about ten language pairs. These translations, available from the home page's Translate tab and the Translate link at the end of every match, make documents throughout the Web accessible to people worldwide, without restriction based on the languages they speak or read. These translations, possible through a partnership with Systran, a leading manufacturer of translation software, have seen incredible popularity since they were introduced on the site.

Franz Guenthner introduced Louis Monier to Dimitrios Sabatakakis, the CEO of Systran, in late August 1997, and the translation site went public on December 10 of that year. As Louis put it, "My kind of schedule."

SOME POINTERS TO OTHER USEFUL CHAPTERS

In this chapter, you've seen how the advent and development of AltaVista were a collaborative effort among AltaVista team members, and how they're truly revolutionary. As growth and development continue, AltaVista will keep offering you revolutionary, cutting-edge features. Until then, you might visit several of the chapters in this book, which show you how to use and benefit from the current features of AltaVista Search.

- See Chapter 2 to learn to use AltaVista's Simple Search.
- Breeze through Chapter 3 to learn about AltaVista's people and business search capabilities.
- Stop by Chapter 4 to learn how to use AltaVista's awesome Refine feature.
- See Chapter 5 to learn about using Advanced Search.
- Visit Chapter 6 to find out about doing Simple and Advanced Usenet searches.
- Go to Chapter 7 to find out how to customize AltaVista.
- Look at Chapter 8 to find out how to most effectively provide information.
- Thumb through Chapter 9 to find sample Simple and Advanced searches.

A Snapshot of AltaVista

The following section shows you a snapshot from a few seconds of AltaVista—that is, all the queries that were submitted to AltaVista within just a few seconds. This is a fairly random listing of sequential AltaVista queries that were submitted at a rate of about 300 per second. The variety of languages represented on this list demonstrates the worldwide use of AltaVista. You'll also notice a number of misspellings and typos—searches done in foreign languages show up with odd misspellings in this list, but they are not all errors. Some, like N|rnberg for Nürnberg, show up with odd character substitutions because the AltaVista log files aren't designed to preserve all characters. However, as you'll notice from the queries in English, not everyone who searches with AltaVista can spell, either!

This list is interesting because it gives you a sense of just what people are searching for with AltaVista—and that turns out to be anything and everything they can think of! Read on to see a sample of just a few seconds' worth of queries…

"screen saver"	pablo neruda	Photoshop and Plugins	vicerrectorados de alumnos
giardinaggio	"xserver emulator for win95"	programming	Astronomie and Kometen
"City Map"	Pittsburg Penguins	Apollo 11H#	xact
+oncnd'	triage and disaster and larrey	miss universe	sed pattern regular expression
+Smit +holland	+"doom2" +gamez	+biotronic	fenwick and fishing and tackle and iron and thread
http://www.rcs.com.sg/p10	"map of Panama"	olyckor med barn	ericsson

handelsbanken	welding nickel alloys	lo	Chat
serenity	technology	bcrazy	J.D. and Powers
"Elle MacPherson"	compact and paper and united kingdom and industry and drying and	"The Simpsons"	opengl/screen saver download
(infrared or ir)	Skdrgerd	screen +saver +parachute	manga
"DECADES" AND planning	+lancia +"gamma coupe"	fosforo	http://www.icom.co.jp/
california cages	bj?rn ruud	Subjekt* and Bildung	IONS IONIZATION IONIZER
auction	Foerderung AND ISDN	+obligationer +aktier	elementary+linear+algebra
weatherstrip	horoscope& taurus	asia+7th century+map	serial number and parasite
laika AND model	lindsay cullen	webpublisher	StrongARM NetBSD
solva	berlinguer	basketball	spain weather
Mdklare	thalia	"australian preferences"	speed kills
+diablo	COBOL+ VSAM	Carrie WEBB	+lancia + "gamma coupe"
aftonbladet+chat	"artikel basen"	Greek Culture	DEC
"Jay Kersting"	"home improvement"	catalina aeroplane	planner and travel
diablo	thunder gt helsinki	hewlet	Luukkanen
Madison and Simmons	peptidase AND lactococcus	"singing voice synthesis"	"Trio DataFax"

A

"skellefte kraft"	Collections Gallery	vidio url:.fr + constructeurs + constructeur + materiels	Orwell
MRB ok~q	"sharp corporation"	Distributor Kingston in singapore	+medical + hospital + wastes
+legislative + francaise + election	guide to online service	"kawasaki"	Illinois mcdonald"s restaurant
"kidpower"	"http port"	+seattle + frequencies radio –climate	Proxy and FAQ
ninfetas	"pictures of cakes"	wrestling	survival game
configuration management	jdrvenpddn urheiluampujat	"terapeutic" touch*	verona
compact AND disc AND accessories	+chess +PC	course X.25	"x3m"+"music"+ "acid"
"turisme de barcelona"	DELCAM International	netscape	korean company in austria
"computer telephony integration"	"windows network"	Smoking	theater
winmail.dat	handwritten	Monash university	David Duchovny
EU-Narkotika-Sverige	paratu	TIME	optical flow
online and books and suspense	Tele1000	"xpower"	Junkers
utah state university	nouvelles de geneve	sed and dos and not unix and not linux and not statistics	free pictures
mp3	laser& triangulation	dna database	telephone AND dial-back

free;games	Conversion near CGM	kevlar	mount everest
Feuerwehr-Fest	Token ring	areotech	midnattsol+golf
IRDA	"Europdische Kommission"	mckinsey	iowa falls, iowa
Doctors	grants	campagnolo veloce	"USS Arkansas CGN-41"
Census	aerosmith	onload	s.r. rangarnathan
"nokia 3110"	motorcicle helmet	"artikelbasen"	smear test
luminance contrast meter	url:microsoft AND powertoys	"group Life"	Holzhandel
"Hiroshima"	Kosta Boda address	idf	"Turbocharger"
cystography	social work	nematode	soundgarden
Scripts AND Shell AND Mail AND Subject AND Unix AND Solaris	ibm voicetype	microbrewer* and "south america"	+institute+ advanced+studies
coding convention for c	www.geocities. com/siliconvally /1080/nasacams. html/	tuinen	+Droege +comp
"Olympique de Marseille"	chambord royale	schack	montgomery and 1944
china newspaper	stock photo online	newsserver free firewall	exhibition fixtures supplier
las vegas	MARS ATTACK	file maker	"John Rowlands"
uudecode	microcode	maldive	skador inom sjdlvfvrsvar
fiat topolino	+kostenlos + homepage	tree;diagram;tool	sms

level round classification	gamez +appz + warez + direct + passwords	"axxion or"	onkologie
codes + mdk	esane	opgavecentralen host:.dk	Skdrgerd
web&database	hermann nitsch	"australian preference"	virtuell ip-adress
"Suid Afrika"	ADSI	India map	Photoshop and Plugins
rsh daemon	librerie and universit‘	british airways	jhmn
handicap	BEIJING	econa101	A friend , Ganesh
picture publisher plus	Pieck Anton	indice ciclismo	METRO AND GOLDWIN AND MAYER
msaccess	+hexagraph + rod*	"cowboy way"	Cgi4
oracle +jobs developer + colorado	Victoria Gas Exchange	Meteorologische Anstalt	gif construction set
CME	computer viruses	Olympiastadion	windows and free and "share modem"
sabena time table	gvhd	foreign currency	"salor moon"
BRUN ELIO	+Einkaufen + Mvbel	aaron k	host:home. netscape.com and Linux and Navigator
Webpublisher	droit du travail	quadram	sharp zr
Berlinermauer or + or DDR	forte agent	New industrial technology	Debt Deflation Hyperinflation
ADB	L+evitov Leonid	Pompeii + volcano	vgetty

Tele 1000	palladin press	cleborne maddux	+vehicle + Germany +cars + markets +motors
"rap"	what os hardware is my unix box running	"Michael Glenn Harvey"	"25cm"
resebyre	APC + schematics	Mwave Drivers	+palmsprings + springs
macintosh utilities	castigo	+Newton + messagepad + board +Canada + install	Chat
plasma thesis dissertation	hidrometeorolo	Window+modem	hewlet
pace university Varden	www. chromosoma	George Orwell	"the guardian"
CATALOGO FILM FESTIVAL CANNES 1997	"Blakes+7"	Carrie WEBB	diana
Tiepolo	trabajador cuenta propia	sonic & knuckles full version	games
+matlab+download	linear programming	"windows 97"	+lancia + "gamma sedan"
"apple daily"	Meta di Sorrento Giosue' a Mare	"linda lampenius"	advocaten
ga lottery	Tegneserier	"randy hansen " guitarist	pirate ship
"jamie lee curtis"	BDP	"courtney cox"	"miss fitness" + 1996

mathcad mechanical	Powerboats for sale + yachts + inboard + boating	tlz7l-ga host:.digital. -host:.digital.net -host:.digital-cafe.com	-host:.us-digital.com -host:.digital-emerald.com
orbit AND records	balonmano	simpson avi pics and sounds	sydkorea
mathematical + symbol +hyper + text +markup + language	directx	voice recognition	"YAHOO"
solstice	+persische + kultur	"education singapore"	"Cirrus Logic 5429 Drivers"
astrid lindgren	instabilities	Luukkanen	Canada+Belarus
dow jones industrial 30	Tammbi	online and books and suspense	FH884k@e0f
+Precipitation -precipitation -rainfall -climate –temperature	-temperatures -weather -forecast -forecasts -wind –moisture	-evaporation -surface -soil -vegetation -snowfall -rain -snow –flood	-flooding -floods -drought -droughts -seasonal
testosteron	intranet	falkenberg	fertilization zygote
marrige thailand	turbo pascal	Media traning	diff lube
"amara ann dunae	Agent+scully	framemaker	novatec sport sedan
Illinois Architectural Barriers Act	+constitutional +convention +host:au -host:republic.org	norcross footware inc.	"the guardian"
+"la salette" +apparitions	software spectrum	"ilections ligislatives"	Calyx

router+software+ download	pregnenolone and depression	Vaughan Public Libraries Administration	persian cats
A_<< =E:P	+hexagraph + rod*	Imppa's homepage	Sun Yu-li Sculptures
Pulkkila	+ferrovie + trasporti + trasporto	+bank socially responsible + funds + investing + savings + checking	+fees +interest
Motorbvrsen	+diablo	duke3d +build + map* +util*	london tourist
numerical recipes	deutsche bank	maintenance AND schedul AND program	Java AND PDF
helsinki	"Golfskor"	book printers	"The Wonder Years"
thailand	supermarine spitfire	seungheelee	"coconut candy"
karlovy vary	randy mamola	standard and Ethernet and IEEE and system and network	+"phantom records" + australia
pclockout hack	USS Missouri	survival game	amiga kickstart.rom
free games	Gillian Anderson	greece and ferries	DAKO Machinery
krone;salzburg	camping AND Krakow	TRAK Microwave	HYPERTENSION
terrance+matron	huge	"denali national park"	lexmark
china newspaper	price	sportboats	Travel Netherlands
"City Map"	Rage Against the Machine	+Droege +comp	"windows networking"

pendal greetings	+Levitov Leonid	+lancia + "gamma berlina"	peptidase AND lactococcus
Roskilde	totocalcio	webcrawl	"Universite Paul Sabatier" & "liste des DESS"
+genealogy + shareware +SQL	+mercedes +pagode	AIM	wwf
win a lawnmower	Rotary	Durette	GQ1[FyF.
"nykvpings musik"	+pkz204* + pkunzip*	+"Flying Dutchman" +xxx	wavemetrics
baney	Gold options	internet+fax+ server	Orwell
Skdrgerd	Cgi4vb	Game Cheating	krone;salzburg
bike-in cinema	Singaporean sculptor Sun Yu-li	OR Technology	timber AND ridge AND lodge AND mark AND twain AND lake
petrified AND Macintosh	MALT+ WHISKY	quantum	vuego acer
"home improvement"	"configuring internet explorer"	Counter+CGI	+distributor + fine +food + gourmet + importer
nasa hubble telescope photos	black	hun yellow	psychic tricks secrets con debunking
autocad	netscape	Douglas+ Edward+ Cummings	Advice UPS
s.r. rangarnathan	ldkare	fruit and spray	whos who
petrarch	Madison and Simmons	bahnboostar	sca
filosofia	Romeo & Juliet	piantina+di+bari	china
pirch32	windows 95 wallpaper	gt helsinki	lumbar fracture

emtex pc	guide to online internet servie	"YAHOO"	"kidpower"
Lenekonvolutt	baby	"Lugano"	solva data
Amazon	heikki+turunen	garfeild	mIRC
deejay promotion	"restaurant industry"+ "press releases" + quarter + revenues	'Happy Birthday'	unix mailing list
mimmi;gulliksen	econ arena	cla na samochody	windows95 character set tables
http servers	fs-on-line	EU-Narkotika	for sale+fish
doc to tif	+"IS-101"	+winpack +zip	"design pattern" AND "object oriented" AND catalog AND online
BRUN	http://www.worldprofit.com/money/macourt.htm	L'Arc de Triomphe"	KLIEN MOUNTAIN BIKES
Behavioural Treatment of drug abuse	transplants kidneys	+education +scan +testing + software	legal terminology
Feuerwehr-Termine	audio speaker+design	Lappeenranta Paper Mill	Powervideo
guide to online internet service	Institut de Mecanique de Grenoble	the beatles	"easy life products"
Mars-Computer-Industrial	Photoshop and Plugins	kavasaki ltd	serial number and parasite
9L=:?y5e	second hand clothing	"who is rfc"	+chess +PC + game

hyattH#EZ	Cyprus + Capital	+"atomic clock"+ cesium +time	BEIJING
monoid	Beautyway Publications	+rcr +std-028	Favereau
Hennef	+"awk manual"	DPCM AND paper	paginacion
win95 casino games	china	+"satellite launch" +hughes	"garlic"
welding nickel alloys	Poskanzer	target	IrLAP
sim-lock	svenskt spel	+palmsprings + springs	political attitudes and china +political + china +beijing + chinese
+hong +kong +attitudes +attitude + sociological + psychological	online and books and suspense	uisl	DAKO Machinery
Cgi vb	monsanto	vapenexport	"tda1950"
"fsis food additives"	political near party near act	+plugin photoshop aiff*.8bi	etherexpress 16 driver
the spice girls victoria and mel c	grolsch	women AND network	sasha
dropstuff	"nino rota"	San Sebastian	restonomi
"amara dunae	richard's realm	bourbon roses	"Paprican"
sparbanken boendekalkyl	Newspapers	giant model airplane	Manchester+ United+Juhninio
Fs0f1b	sodiumchlorate	file maker	spagna
volna+mista	Rush MP3	"laser diode " AND driver AND NOT avtech	s3 3d
soundstream	gun ballistics	listino alpha	bvrse

Felicity Murphy	Micro- Loan	birthday	midnattsol+ vildliv
"two voices"	FK<[	"fuel" near "water" near "contamination"	chainmail
lexmark	+Droege + comp	harga and mobil and bekas	+Newton + messagepad + accelerator + install
andrea boglione marino	brave AND heart+ AND pictures	"Jessie Bannon" (From The Real adventures of Jonny Quest	arj
software spectrum	+hawaii +real + estate + photos	Troax Axo AB	where can one get song lyrics by searching
mpeg movie free	tcp_rexmit _interval_min	"miromedia online"	tore eriksson stockholm
Suicide	New Mexico boys ranch inc.	+sondage	Distributor Kingston in singapore
Huge	"meteorological office"	exco	"easy life products"
CELTA DE VIGO	+deskjet +500c + printer + driver	luonto	"private investigations" florida tallahassee edwards A9700076
gif sport	+"intellimouse driver"	+control mechanisms + change + education	beatles 1962
consul	hong kong culture export	protect and mailing and bomb and secure and script	skiing

pirates	ericsson	baglioni	"Janis Joplin"
"configuring internet explorer"	"ona zee"	Bornholm Tourist	Modern Art
jonny quest	anonymail	Petit Blanc	U2
Rauenstein	Real Club Celta .Plantilla	+DCT +cosine description overview introduction	dna library
radio frequency research software	alta vista	serial number and parasite	Ac/dc guitar tabs
puls*tv2 host:.dk	emballage	anorexia	high & frequency & filter & crystal
object-oriented&sof tware	beth AND lisick	Rockhaus	sodium chlorate
sound deck	Perl	amiga kickstart.rom	PNS
"vdsteres graffiti page"	emaildirectories	"restaurant industry"+"press releases" + quarter + revenues	arrowtech
Debt Deflation Hyperinflation	"women's movement"	token sales	dow jones industrial 30
'MOL' AND 'Szolnok'	Versicherungsve rtreter	vicerrectorados de alumnos	Aromatherapy
digital card	"hizkuntza"	?;A&CC	Canadian submarine links
corel	"tariffe telecom"	ancient greek buildings	Cipriani
sharp zr	Nisses Hemsida	orbit & calculation	montgomery and 1944
IrLAP	+persische + kultur	grants	Par'
harmon, bernadette	necsy	Hauppauge	jonny quest

"Internet Explorer"+ "Language Pack"	shearer pitcher	fxbnd861	MALT+WHISKY
black women	Smoking	+"computer telephony integration"	A6<10K
solva,data	Calyx	midnattsol	Twins Magazine
"education singapore"	Hiroshima+ map	A-team	directx
ICS 1700	device driver catalogs	Barclays Global	Musik
stomper +isdn	hewlet home	Luc Besson	Proxy and FAQ
finger	+ibm +2456 + "announce letter"	toilet draining slowly	Madison and Simmons
Terrorist+Handbook	bahn-booster	Chatting	Manoj Bahl
Huge	hardcore	design	"GARMIN"
link:http://www. alpine.com.au - host:http://www. alpine.com.au	protect and mailing and bomb and secure and script	nodefaultlib	+Droege +comp
JAVA	elektronik	+ledarutveckling + ledarskap + konsult + utbildningsprogram	livemen
VRML	"love advice"	telephone	newcastle soccer team
goran ivanisevic	+Bill of + Rights	social work	survival game
spice girls	9q8#.a	giro italia	POW WOW
Skdrgerd	deejay promotion	football coaching chat	Collections Gallery

air+national+guard	+whole + number + numbers + fractions + decimals + subtraction	+multiplication	french and elections and polls
"25cm"	databvcker	entera	pendal
gt helsinki	cobol	nekton	resebyre
timber AND ridge AND lodge AND mark AND twain AND lake	profezia di celestino	tomb raider psx	Holzhandel
clipart	"live to ride"	cobol mvs pc	consejo colombiano de seguridad
ultimate.zip	fortran77 sources read oounit open	ERWin	ericksson
U2	"Jas-39"	boote	"anestesia-rianimazione"
+hoppenstedt + domain:de	windows board game	design employment, ma	give me fire wall
topline	+Livorno + Palermo + Poltrona	sebanken	gordon conference 1997
westwood	Fw=:EM hyperlink mailto:5p@Z@N 5p@Z@N	imap exmh	+genealogy + shareware
"Asian development bank"	diff lube	norcom AND satellite	diablo ver 1.03
marrige thailand	ultuna	"golf clubs"	volvo
+merkaba + spiritualist + interdimensional + ascension	SUNG HI LEE	"kidpower"	peptidase AND lactococcus

sound + player	ducati accessories	testosteron	Bk'dhlhp Jsgllhm
"Paprican"	windows95 register problem	playstation	"Daggerfall" + game
mian mohammad nawaz sharif	goal storm	peter messa	powervideo
"Legends of Hollywood"	coke plant	china+7th century+map	rg62au
"cypress college" and councellor	celebrities	brodrene dahl vvs	Hvchstleistungsrechenzentrum
+ALPS + "keyboard layout"	Audi near Germany	a+	iso
"Kimberlee Acquaro"	+canine + leishmaniasis	+Droege +comp	what is database
cloning	Bjvrn Larsson Norstedts	glass association in singapore	a-10 fairchild
canadian newspaper and toronto and business	previsioni meteo	"dallas mopar"	steelhead
+Wenzel +Sedlak	soraya	"Finnair"	+networking AND "switching hubs"
modem user manual	++rencontre agence	Zara White	50 AND rheinland
3d graphics cards	history essays	+Newton + messagepad + accelerator + install	Kissing + disease
jeti+kip	NT service pack 2	small business	polskie ksiazki telefoniczne
BEIJING	simpsons	nakamura beri osaka	sca
kelly blue book	chips italy	AxHo?U	internet service finder

bvrse	+sondage	etherexpress 16 driver	finance
bahn booster	e.m.remarque	"acer peripherals"	"HP DeskJet 870" AND "NT 4.0" AND "printer driver"
texas AND census AND index AND and AND starr	china+7th century+map	modrow	+6x86
verona	kxbenhavn host:.dk	exite	Jeep
brigitte nielsen	english and grammar and educ and online	Bronco AND Burger	women AND network AND Netherlands
blood AND pressure	MICROSOFT +WORD	xact and scilab	"convert icon"
"home improvement"	win32s130	"halifax building society"	plasma +thesis + dissertation
"xpower"	X0208	Monash university	mcafee and update
"uppsats om snusets skadeverkningar"	suchmaschine	protect and mailing and bomb and secure and script	+education + scan +testing + software
multi-x	Hovedbane-gerden	dec_ggcf.zip	free jpeg men
audio speaker+design	Illinois Architectural Barriers Act	+"satellite launch" +hughes	BDP
pregnenolone and depression	Cgi vb	survival game	www.leisureplan.com
"mvrk materia"	+windows95 + faq +help	lumbar fracture	+embedded* + system* + development* + platform*

rfsl and webchat	china development	mpumalanga	china+7th century+map
kosovo	sonnenenergie+ wdrmepumpen	Pilotrite	file maker
Debt Deflation Hyperinflation	"Rover 420"	welding nickel alloys	ultram
Serenity	vgetty	Perl + (uni bayreuth)	;g@|
differences american australian	sipadan	matsumoto-megumi mamegu	prosolva
cheats_to_"z"	"cause+and+ effect"	"modern art"	lexikon
maru AND 12 AND font AND .jp/ AND X11	"Adam Wolfsdorf"	juliet	kavasaki ltd
xian	beavis butthead	corel and cd and creator	+persische + kultur
"anestesia-rianimazione"	netbewi	"magic the gathering"	"banking passport"
link:paranoia.com/ faq/prostitution	"everclear" + "guitar tabs"	Coka-Cola	Rubik's Cube
"roderich bott"	tim	Chat	programming AND threads
+Shaw+High+ School+Reunions	lauritzen host:altavista.so ftware.digital. com	telephone service	10310 Brockwood
www.se eu alkohol sverige vin	miko and frank	asphyxia	ADSL
www.sr.se	morph pc	bellevue, Washington	Carrie WEBB
hyperlink mailto:32@Z	3D Data	Swiss Air	FIS

massage+cushion+ for+car	western australian classifieds	vinyl search	Gold commodityoptions
michael biddle sydney australia	krone	+"computer telephony integration"	(AWP AND game AND music) AND NOT Washington
+orientering + Oslo +bedrift	Distributor Kingston in singapore	air+national+ guard	opetussuunnitelma
Skdrgerd	+osiris + software	skinhead ry	killer
richmond high, Richmond B.C.	+ibm +2456 + homepage	elementary+ linear+algebra	haustaubmilbe
shipbuilding	"LED DISPLAY"	how to dupe items clone diablo	Institut de Mecanique de Grenoble
boote	Giubileo	"oregon state university"	regeneration+planaria
volna+mista	wwf raw	A6<10K	analog devices
"DECADES" AND planning	cobol +mvs + pc	anna +horny rob	tribune and de and geneve
Grants	nursing AND ethics	+pkz204* + pkunzip*	formula1 benetton
Ufo	Elvis	impimis hard drive	universit‘
macintosh hacks	wal-mart AND job AND opportunities	huge	political near party near act
Ax7JA_GP13	Scalise	+simm +"pin" + 72 +sheet	Win*95&logo
address AND of AND prentice AND hall	dam+fotboll	consul & Mishra	+diablo

security	+"Flying Dutchman" + xxx	PLANTILLA EQUIPO CELTA, RATKOVIC	A friend , Ganesh in Terengganu
C5>H=C	ankylosing spondylitis	NHL teams	flowers
book printers	exhibition fixtures supplier	"Lugano"	Industrial+ Reverse+Osmosis +Systems
+gamez warez	Ax7JA_GP13	games	cookie
protect and mailing and bomb and secure and script	cam_data.c for EXB-8900	link:paranoia.com /faq/prostitution	process engineering software
bracket	"X-Files"	lords	Marijuana (Against)
gt helsinki	Logo	algemeen rentefonds	front242 AND headhunter
cd and music and on-line	mazda mx3 parts	fleming england	audio visual traning
Supercruiser bus	DAYS INN LISTINGS FOR IOWA	canal+plus	mangas and shoes
university english	certosa di pavia	cranberries	"electronic telegraph"
"education singapore"	"narrow way company"	statskontoret	pendal
win95 casino games	Evangelion	+Realmz + Download	sharp zr
"Paprican"	HDB Singapore	"omni-nfs" AND "serial" AND "number"	+LAN +lan + wans
euroll	free english translator	hvad er xkologi host:.dk	sipadan
Hubble	+the + cranberries + music	keeping+fit	+"Richard Axel"

Ax7JA_GP13	+HTTP + forms +GET + tutorial	Madison and Simmons	Kosta Boda address
architecture	chaos	sound deck	gif and sport
vicerrectorados de alumnos	Micro- Loan nevada	"michel jordan"	ghostview+win
Myeloperoxidase	mediamagic + homepage	Sarenna Lee	"design pattern" AND "object oriented" AND catalog AND online
dawn whiteman	Australian Newspapers	sex+pistols+ pictures	listino alpha
Bedford School	http://www.pingweb.de/celeb/mnamsort.htm	netbew	"ericsson telecom"
linear programming	oscar	finance+securities +risk	"ona zee"
+education + control mechanisms	italy map	free download	"radio broadcasting" + AM
startuplogo nadesico	DIARI OFICIAL DE LA GENERALITAT DE CATALUNYA	reise	Keitges
bahn-boostar	KLIEN MOUNTAIN BIKES	Ax7JA_GP13	"halifax building society"
databvcker	petrified AND Macintosh	utah state university	9743

ticket printer	+warez + gamez + theme + hospital	lemonheads	windows and free and "share modem"
hotel romantic sitges	doc tif	social marketing	Aalborg-biograf
causeway + new arleans	"Legends of Hollywood"	census	"birth records australia"
tjava	sca	internet AND millionaires	doctors
midnattsol	"More beautiful site"	humidity AND indoor AND typical	Ax7J A_GP13
olives	otranto turismo	20th century fox film	vicerrectorados de alumnos
+Quick +View +Plus +serial	moottoripyvrdn liikenneva-kuutus	+"intel triton" +ide +drivers	"ntcrack"
la tribune de geneve	lords	+constitutional + convention + host:au - host:republic.org	washing AND machine AND washer AND fuzzy
"Call of the Wild"+Jack London	"rohrpanele"	etherworks nt driver host:.digital. -host:.digital.net	-host:.digital-cafe.com -host:.us-digital.com
-host:.digital-emerald.com	Mafia	"Exeter Devon"	filemaker
"Febuary 8th"	twolines	+bridge +spiel + demo	montgomery and 1944 and war
+miniblinds	"HP DeskJet" AND "NT 4.0" AND "printer driver"	hewlet home	blue poster

Universite Catholique de Louvain	seagate compression	gods word for windows	ericksson
Crawfish	"Formac"	Belarus+Canada	bvrse
"Rise of the Triad"+"download"	free cgi scripts	jason voorhees	roses
AxHo?U	toilet	scriptie + EDI	"survivor benefit"
(marketing model)	a+	pirates	"nature scenes"
Bermeja	"xpower"	BRUN	litchfield
Huge	"YAHOO"	orbit & calculation	bodhan
Resebyre	usr pilot sofware	pics of geri	"tariffe telecom"
quadrennial defense review may 1997	star trek+desktops	pesca	EP600
WalMart	sun.java.com	call waiting how to build you own curcuit	handla AND hemkvp
amiga kickstart.rom	+Saguenay+ Bed+and+ breakfest	servodan	web&database
C-64 AND kaufe OR C-64 AND suche	+Lego "Zbigniew Libera"	"25cm"	k6 prices
Fossnes	previsioni meteo	ERWin	command
JAVA Scripts	proso	reskit.exe	u2
New Zealand	+"computer telephony integration"	Herbie or Herbert Lee	holidays
BEIJING	french and elections and polls	clipart mailman	social marketing

women AND network AND Netherlands AND education	+Landis +Gyr	hausstaubmilbe	Mirc
schumpeter	dtstvrningar+ anorexia	Nisses+Hemsida	yle radiox
'Happy Birthday'	furniture	"9L=: 4YC#4O"	windows95 character set
index+server	survival game	"condivisione di indirizzo" AND software	Cyndi Crawford
1997 AND nba AND draft AND prospects	an	Skdrgerd	differences american australian
sasha	rcs	(marketing model)	Twins Magazine
"litigation support"	elementary+ linear+algebra	+perl +rdb +free	Cgi vb
diff lube	jo guest	+CARLOS + KLEIBER	etherexpress 16 driver
Pilot rite	jisx0208	ungdomsprojekt	yatsura
Utbildningsbidrag	(University of Buenos Aires)	"vinnare pe bingo lotto"	Trovatutto
english and grammar and educ and online	Goldman+ bilder	Midland Liquidators	"home page"
telecomandi and remote and control	+ibm +2456 +homepage	pregnenolone and depression	USPA
grants	mpu 401 programming +mpu + soundblaster + midi + makedev – joystick	-scsi -ide -mitsumi -atapi -rom -tropez -turtle -synthesis –sampling	-distortion -beach -tahiti -harmonic -roland -voyetra -orchestrator

-etherworks -pci -cyclades -cyclom -smc –adaptors	what is the story morning	Drug legalization	Debt Deflation Hyperinflation
Cindy Crawford	ung i EU	"boot list"	vulcanoes
+percorsi +stradali –adria	amphetamine manufacture	solna	POVRAY
jackey chan	hotels, The Chelsea, London	+Linux +16650	regeneration
"star seekers"	ultuna+hemsida	source NEAR routing	UPMC215 and Light Top
prodigy	opengl/ screensaver/ download	visual c 5	"kidpower"
hunting	"Food Channel"	restless+modern+ jude the+obscure	"duke_nukem_ 3d" +"download" +full_version"
Skdrgerd	soundgarden	kesdmvkit	uni-stuttgart
Industrial+Reverse +Osmosis+Systems	huge	thebest- doctorsinamerica	"xpower"
+tamagocchi + "6 times"	The Washington Monument	mont blanc	consul & Mishra
startuplogo	"www-server" + Erkldrung	"bridgette monroe" and not excal	www.goe.com/ daily.html
tribune and de and geneve	kongehuset+ Danmark	kastigo	pthreads + linux real time -decthreads
warez and helpmaker	Aftenposten	downset	Refused
ieee	"strauss family"	Industri senteret	CANARIA
kristina fren duvemela	Ax7JA_GP13	Gold commodity options	keeping+fit

internet AND millionaires	ces	"States with Runaway Program"	capasitif proximity switch
distributor	+christian+ henningsson	"map" + "scandinavia"	coal AND warratah
bombshell	"time-lines"	+cccp +"fedeli alla linea"	monoid
sun coast australia	ghostscript and windows	+Christoph + Breiter	koreameeting
melonville	Roulette	+canine + leishmaniasis	zotos
opale	magiccom	REAL CLUB CELTA	skolarbete socialism
"bimetallic catalyst"	flesh eating infection	Pioneer	Stunt Coordinator
COBOL+VSAM	Packard Financial Hewlett	flowers	foreign currency
+diablo	mange	technology management strategy	erotic and screensavers
"Corbin Harris"	divola	" HIGH SPEED SERIAL INTERFACE " HSSI	Skdrgerd
Clean Sweep	patent classes	hart	"education singapore"
Stray Cats	shirley bassey lyrics	survival game	linux AND HP AND printer AND laserjet
acoustics and drum	harden beat	cobol +pc + shareware	United Church of God
patrick crispen +crispen +roadmap + workshop +crispy + listserv	auto hause verlag gmbh	"Parupudi"	London bank

koi	western bulldogs	"Paprican"	ez3200p+ driver update software
pesca	++rencontre agence	Stunt Coordinator	osprey aircraft
conan obrien guest	"kogan publishers"	ovaneker	"Mariko Morikawa"
Eurostat	+"atomic clock" + cesium +time	boote	+introduction
+johnny +wahlstrvm +piano	claris and homepage and demo	"steve jobs" and apple and newton	Evangelion
florianopolis	mesar	+presentation + "search engine"	Stunt Coordinator
Ek=EFG8E	planers and planer moulders	give me fire wall	bahn boostar
Skdrgerd	sound deck	Stunt Coordinator	koordinierungs AND beratungsstelle AND bundesregierung AND
Informationstechnik	explorer australia	Madison and Simmons	DEC
Proso	(long near term or longterm) near weather near forecast and russia	hanson	buzzerbeater
Startrek	+Fremdenverkehr +Nordsee +Schleswig	midi+sanvoisen	animations
S.D. COMPOSTELA	WHIPLASH	malaria	yle radiox

Mercuri	+host:austria + host:at + wertpapier	classical cd's	"rap DRIVER"
swing detective	goal storm 97	"aol"	"home page"+calris
marrige+thailand	california cages	america	political near party near act
Sca	marilyn manson	la tribune de geneve	mr.president+mp3
Sarek	5e82=: @NEM>WF<:j	netscape+linux	protect and mailing and bomb and secure and script
switch* near video near (vga or svga)	michael biddle sydney australia	"The Queen City Of The West"	nematode
+GQ1[+@NEM3]+ 0K;v+?#Ax	michaelbolton	"firewall configurations"	TOAHIBA CD-ROM XM-5602B+ NetWare 4.11
+simm +"pin" +72 + datasheet	canon printers	+Debt + Deflation + Hyperinflation	Distributor Kingston in singapore
'Heavy Metal' & '.mid'	krone;zeitung	emerson+lake+ palmer	"marilyn manson" AND photos
COBRA	gold coast	WWF The Arcade Game	NPR Initiatives
Nederland; weather;statistic	hewlet home	credit card lists	link:www.dsm. org
spa controls	Hennef	formosa +boat	instabilities
+microsoft + explorer +internet + software + download + downloading	toilet	call waiting how to build you own curcuit	"Legends of Hollywood"

westwood	giro and italia and 1997	camping AND Krakow	tennis
Powerboats for sale +yachts + inboard +boating	f-prot	MOONEY & nonfatal	paracaidismo
Immobilien	"Micronics W6-Li" "Phoenix BIOS"	http//:www. maklarc-etuna.se	Napoleon
fuji AND provia	6000Video	TrueSpeed	toni braxton
vidio url:.fr +constructeurs + constructeur + materiels	9th Inter-Service/Industry Training Systems and Education Conference	Holzhandel	archeology
beach and babes	Cyndi Crawford	card reader microchannel	coulter+ tranquillity
+iran +pictures	Shomer-Tec, Inc.	pomona	online subscription mailing list
Ungania	"PEOPLE SEARCH"	huge	9P85
Sasha	information need journalist broadcasting	nsr group	warez
simmons wheels	Cgi vb	Conversion near CGM	mel c i spice girls
+ibm +2456 + homepage + scanner	furniture wood	"bingolotto"	merikarttanavigointi
reference letter	PLANTILLA EQUIPO CELTA, RATKOVIC	mexico AND cars	+hoppenstedt + verlag

"volvo"	+sondage	sung hi lee	autocad
iisalmi	Xircom+ 10/100+ Ethernet III	duncan dhu	"ona zee"
http://www.rcs. com.sg/p10	anabolic steroids in body building	"claudia christian"	Document Storage
proso	free cgi scripts	"notes vs intranet"	nlrnberg +flughafen + dr.schwarz
palace and brothers	=aimslab tr288 +linux	Herbie or Herbert Lee	cross olours

What People Are Searching For

The following lists provide a glimpse of some of the things that people search for using AltaVista. These lists were not scientifically calculated and are not intended to provide comprehensive lists of terms within each category. Instead, they represent some of the most searched-for terms and categories from the top 2,000 most commonly searched-for terms. Think of these lists as a guide to how diverse AltaVista searching can be—and as a source for AltaVista search trivia!

Top Colors

Black	Blue
Red	White
Yellow	Brown
Green	Silver

Top Company Names

Microsoft	IBM
Disney	Sun
Sony	Yamaha
Packard Bell	Compaq
Intel	Oracle
Adobe	Epson
Lotus	Amiga
Hewlett (Packard)	Motorola
Toshiba	Panasonic
Netscape	

Top Computer Programs/Names

C	Word
Win95	Macintosh
Windows 95	Windows NT

Photoshop	WinZip
Doom	Pentium
Eudora	Visual (Basic)
Java	Pascal
HotMail	PostScript

Top Computer Terms

www	http	post
gif	jpg	software
download	chat	games
home	net	computer
internet	world	web
driver	image	html
files	mp3	mail
home page	midi	online
sound	shareware	server
java	link	alt
design	url	information
network	images	avi
list	code	memory
pc	ftp	zip
linux	wave	mpeg
pic	modem	technology
program	sites	3d
exe	virtual	virus
file	directory	serial
dos	unix	programming
database	htm	printer
freeware	basic	clipart
archive	cd	computers
drive	disk	screensaver
software	domain	animated

tutorial	scanner	wallpaper
cgi	fonts	fax
port	manual	format
telnet	multimedia	hardware
downloads	ip	scsi
cdrom	mov	conversion
bios	mpg	motherboard
technologies	servers	remote
publishing	command	scripts
converter	chat	interactive
javascript	icon	version
application	tcp	antivirus
hacker	ethernet	network
compiler	cyber	protocol
micro	gateway	error
usenet	isdn	scan
filter	newsgroup	problem
id	os	

Top Domains

com	de	net
uk	us	edu
se	at	it
en	st	no
ca	my	gov
au	dk	nl
jp	be	es
fi	ac	fR

Top Countries

America/United States	Canada
Australia	France
Mexico (could be state or country)	Singapore
England	Germany
India	Ireland
Korea (North or South)	Spain
Sweden	Argentina
Vietnam	Russia
Japan	Thailand
New Zealand	Malaysia
Brazil	Indonesia
Taiwan	Scotland
Greece	Holland
China	

Top Leisure/Hobbies

Music	Magazine	Travel
Book	Weather	Golf
Movies	Gallery	Family
Library	Guitar	Hotels
Food	Books	Jokes
Camera	Magic	Island
Lake	Fishing	Genealogy
Television	Kids	Personal
Chess	Mountain	Tools
Story	Stars	Wedding

Social	Reading	Lotto
Band	River	Shows
Fashion	Repair	Airport
Dance	Arts	Wood
Comics	Camping	Poetry
Paint	Party	Creative
Personals	Vacation	Bike
Quotes	Tickets	Shopping
Dating	Flowers	Poems
Culture	Tennis	Boats
Bicycle	Cruise	Poker
Collection	Beer	Race
Resort	Tourist	Plants
Holiday	Friends	Call
Jet	Theater	Concert
Ocean	Budget	Fun
Coffee	Train	Opera

Top Medical-Related Words

Health	Medical	Research
Cancer	Hospital	Disease
Blood	Therapy	Syndrome
Psychology	Fitness	Nursing
Surgery	Weight	Pain
Smoking	Brain	Vision
Aids	Muscle	Cell
Birth	Skin	

Top Names

John	David	Smith
Michael	James	Richard
George	Kelly	Elizabeth

Gillian	Bill	William
Peter	Frank	Jennifer
Lisa	Martin	Patricia
Mary	Max	Linda
Sandra	Jim	Scott
Robert	Ann	Paul
Don	Steve	Charles
Larry	Joe	Alicia
Barbara	Sharon	Marilyn
Sarah	Chris	

Top Pets/Animals

Dog	Cat	Horse
Fish	Dragon	Mouse
Bird	Pet	Eagle

Top Profession/Education Related Words

School	University
Job	Education
Employment	Management
Research	Federal

Top Sports

Golf	Baseball	Fishing
Wrestling	Football	Soccer
Racing	Hockey	Basketball

Top States

California	New York	Texas
Florida	Washington (DC or state)	Ohio
Carolina (North or South)	Virginia	Colorado

Arizona	Michigan	Georgia
Maryland	Oregon	Wisconsin
Minnesota	Hawaii	Indiana
Maine	Massachusetts	Pennsylvania
Oklahoma	Alaska	Kansas
Tennessee	Utah	Alabama
Connecticut	Missouri	Kentucky
Iowa	Illinois	New Jersey

The Most Common Words Found on the World Wide Web

A list of the most common words on the World Wide Web reflects the academic and computer science origins of the Internet. The top nouns found include information, page, net, home, index, data, system, and university. The numbers shown here are for comparison only—they keep changing (and getting larger) whenever Scooter, the AltaVista Search spider, combs the Web for pages, but the order of the words tends to stay relatively stable.

HOW TO SEARCH MOST EFFECTIVELY

When you use AltaVista Search to find something, it is important to keep in mind that the words listed in the following table are common—meaning they are found in a great number of documents. Including them in your search command may not be very useful, since so many Web pages contain these words. If you search for a term that appears many millions of times, you'll certainly need to add more information to narrow your search. If your searches contain only very common words, your search results will turn up a large number of hits that may not be very useful to you. When you can, it helps to use rare, or at least not-so-common, words in your searches.

Here are some statistics on the number of times the most common words occur in the AltaVista index.

320790448	the	1443597	Tax
179215932	of	1442541	Able
166532454	and	1441297	Resources
153313775	to	1441241	Material
148955456	a	1439247	Either
115242052	html	1439115	Keep
109799540	in	1438117	Terms
101542861	gif	1437655	Mark
98881253	http	1437115	Display
73893892	www	1437101	Manager
69287589	for	1433818	Looking
62271443	is	1433651	Win
61775893	htm	1430943	Doc

61250795	com	1430223	Die
47588960	The	1430092	Section
46472783	on	1426569	basic
42969102	that	1424860	oct
38813093	I	1424781	release
37625741	with	1424319	organization
37203490	A	1423880	energy
35629664	or	1423500	een
35452068	by	1420932	je
34899178	you	1420220	drive
34584917	it	1420002	due
33950297	be	1418477	environmental
33354503	this	1415080	least
32299402	images	1414844	dans
30777250	as	1411891	Jul
30391921	are	1408767	Box
29163142	at	1407746	institute
27033269	from	1407562	professional
22857933	index	1406993	event
20944323	an	1406129	Email
20061487	not	1406117	trade
19974922	was	1404595	materials
19773057	have	1404225	legal
19714126	edu	1404169	means
19453044	all	1403714	che
19029067	will	1402567	wsapi
16789906	1997	1401702	Go
16438976	your	1396707	percent
16425563	we	1393837	fi
16374186	home	1393481	status
15345199	page	1392757	lang=xx
15076587	can	1390172	Server

14966112	la	1385150	faculty
14964531	new	1384618	early
14946512	if	1384398	written
14694522	no	1383189	cc
14359308	jpg	1382577	write
14029513	net	1379841	unknown
13817742	cgi	1379414	done
13609455	mailto	1376638	format
13302001	one	1375655	stock
13247648	information	1375240	resource
13038671	he	1375077	unit
12708630	but	1374665	Washington
12643647	has	1371711	michael
12449021	which	1368845	hot
11913635	1996	1368559	car
11768858	they	1368526	radio
11719209	about	1367626	nov
11594617	other	1365416	West
11524649	more	1365356	Development
11103263	This	1365203	global
11033099	en	1364641	child
10821199	may	1362984	query
10597812	bin	1359368	east
10380687	time	1359054	buttons
10355297	up	1358071	countries
10049560	their	1357608	nasa
9893642	his	1356719	robert
9737613	In	1356394	various
9694687	do	1352855	marketing
9654930	what	1350475	Info
9467658	my	1349878	Feb
9221551	our	1349481	picture

9171113	us	1349186	says
9134722	there	1348704	Music
8965468	re	1347111	reviews
8881161	any	1344992	learning
8539363	when	1343206	archives
8501831	news	1342865	division
8491654	so	1341492	met
8453989	mail	1340471	ma
8423236	who	1339422	short
8413728	also	1338297	America
8405157	out	1337493	Press
8043908	Page	1336811	interface
7906411	web	1336295	result
7773910	back	1336239	pro
7752601	It	1333181	Mar
7749510	use	1332539	lib
7591279	icons	1331940	Public
7581285	search	1331433	High
7547700	Home	1331233	Family
7520832	been	1329509	dos
7516124	data	1327730	Contents
7487921	these	1326549	texas
7399421	were	1325708	Art
7344804	only	1322866	select
7336463	some	1321815	council
7324418	co	1320214	construction
7320738	org	1316786	front
7301350	would	1316531	os
7257001	first	1315944	night
7233202	top	1314475	bit
7108874	system	1314354	Support
7062712	its	1313285	true

7050001	der	1313077	away
7001224	die	1311798	generated
6944445	services	1309929	story
6912009	me	1308840	become
6889010	state	1308807	received
6761397	internet	1307461	copy
6630163	name	1306974	House
6617449	1995	1305733	fact
6615638	graphics	1302481	create
6580425	next	1302234	church
6529611	had	1302113	int
6517420	If	1302017	Products
6494093	university	1302001	paul
6421570	und	1301480	planning
6410772	two	1300544	mac
6396489	how	1300496	om
6320563	business	1299886	posted
6282221	jp	1299279	whether
6238791	people	1298146	october
6227272	than	1298005	war
6222826	New	1297944	method
6182033	et	1295727	collection
6181075	into	1294624	enter
6128051	site	1293651	Here
6055971	list	1291779	third
6000614	info	1291223	months
5980859	help	1290859	district
5978793	year	1290621	dir
5957509	like	1290319	together
5954415	world	1289568	yet
5925516	University	1288701	self
5921795	se	1288457	Jun

5906101	get	1287663	Institute
5896841	We	1286777	Corporation
5890463	service	1286622	Design
5848627	ca	1282922	december
5798091	see	1282201	lines
5772766	date	1282008	client
5734634	des	1281444	designed
5705885	program	1281395	sur
5690907	Re	1281050	update
5666062	number	1280791	sure
5627115	over	1279577	often
5606586	que	1279268	James
5595258	here	1278920	annual
5545842	To	1278717	try
5540105	Internet	1278035	photo
5523930	work	1276490	Apr
5509303	For	1273704	january
5483354	di	1272177	ein
5474617	such	1270898	ltd
5460725	uk	1270886	direct
5455742	her	1267779	Sun
5430103	them	1265916	telephone
5405671	You	1265359	works
5382083	should	1263575	young
5380683	le	1262666	commercial
5354179	software	1261469	ny
5328064	last	1261046	gallery
5282608	inc	1260446	I
5282516	All	1256304	later
5206285	research	1255438	Description
5181812	each	1252889	Some
5179561	company	1251464	Net

5121149	file	1250574	Price
5100458	then	1249837	persons
5089941	now	1249559	chat
5062623	most	1246912	schools
5033210	just	1245257	misc
4992824	img	1243322	eine
4992572	she	1243166	visit
4991512	image	1242173	past
4945875	line	1240377	summary
4938170	go	1240358	records
4918989	support	1240104	publications
4914843	el	1239400	De
4907303	through	1237391	October
4898825	ftp	1237081	economic
4884401	subject	1236977	Directory
4863366	used	1236571	son
4850474	ac	1235736	bank
4784600	un	1232774	william
4780366	THE	1232772	Welcome
4764303	well	1231454	Michael
4760498	yahoo	1230731	Report
4752377	available	1229414	points
4751959	after	1227365	Now
4736553	contact	1226041	having
4730747	public	1225069	far
4729004	No	1224944	Homepage
4725714	fax	1223638	January
4692668	copyright	1223522	estate
4691375	school	1221900	articles
4683278	text	1221570	Engineering
4577671	Web	1221488	Table
4565382	systems	1220311	Oct

4511403	Inc	1219742	limited
4505119	per	1219682	President
4489181	many	1218977	tech
4481262	day	1218209	Address
4463711	make	1217934	pp
4446586	free	1217932	After
4428036	where	1217680	Committee
4413164	products	1217473	JPG
4411678	order	1217275	1989
4390340	Information	1216256	Robert
4376913	general	1216209	discussion
4373316	computer	1212528	among
4332190	links	1204115	knowledge
4300037	under	1202250	original
4294885	group	1201826	However
4287177	said	1201722	natural
4286548	years	1199233	operations
4283178	must	1198978	Project
4265741	high	1196429	words
4259279	return	1192501	print
4222435	gov	1192245	IBM
4210940	au	1189830	Link
4137928	city	1189591	force
4126008	GIF	1189037	dem
4120301	OF	1188657	introduction
4077731	1994	1186474	party
4064358	please	1184325	addition
4052230	very	1183544	weather
4048233	message	1183353	error
4037109	He	1183012	known
4037063	les	1182547	november
4030354	del	1181906	December

4003816	good	1181885	administration
3931799	part	1180333	exchange
3899222	am	1179523	Free
3896737	center	1179161	thomas
3895059	form	1179009	ever
3864447	development	1178935	included
3848400	email	1177243	unix
3826236	national	1175992	Texas
3825328	using	1175777	income
3816009	il	1175320	region
3811436	set	1173998	als
3808089	education	1170998	Dec
3797915	gifs	1168365	meet
3778650	online	1167855	death
3770930	area	1164772	held
3759337	main	1163090	store
3758749	type	1160207	More
3738129	way	1158199	mon
3734868	pages	1157476	Society
3732919	management	1155655	History
3715348	Copyright	1153602	Nov
3713584	students	1151583	Community
3713200	know	1150104	costs
3700764	map	1149913	Paul
3673556	What	1148905	br
3656204	don	1148794	nt
3641866	international	1148645	op
3628262	those	1148616	fall
3588187	access	1147473	fun
3583164	health	1146243	computers
3576001	could	1145393	added
3551899	local	1142883	Do

3539916	address	1141104	conditions
3524269	life	1140674	film
3522633	because	1140166	volume
3513521	art	1139906	structure
3503467	section	1139846	standards
3501439	made	1139743	communication
3483931	title	1138898	Park
3482332	server	1138789	linux
3477475	right	1138570	details
3475192	three	1137915	already
3469751	st	1137766	response
3467030	library	1136710	safety
3461424	before	1136143	taken
3455743	Index	1134846	magazine
3455338	need	1134804	alt
3451722	network	1134510	Club
3434935	design	1133583	Message
3433778	music	1133371	Director
3409755	between	1133298	easy
3398142	office	1133203	along
3357859	course	1132775	hp
3354540	technology	1131981	virtual
3348026	News	1131704	bullet
3343595	du	1130404	father
3341274	From	1128071	qui
3336532	click	1126165	comp
3314320	find	1125146	hotel
3308748	resources	1123437	geocities
3295677	And	1123033	six
3295647	previous	1118150	werden
3292973	comments	1117270	William
3290447	class	1117247	tell

3290124	john	1117000	earth
3284913	both	1116855	employment
3279811	May	1115677	Up
3268429	los	1115547	currently
3249220	him	1114973	Order
3242214	man	1113575	japan
3234062	control	1112660	necessary
3228158	long	1112251	ip
3223156	members	1111916	topics
3221530	State	1111883	although
3214882	department	1111861	documents
3192988	best	1110459	offer
3186554	following	1110261	ask
3174909	windows	1109302	em
3172799	take	1108916	share
3159936	phone	1104598	half
3159506	off	1103908	Use
3155297	same	1103756	lang=ja
3146983	Services	1103476	feel
3133434	den	1102533	ph
3125906	rights	1102078	amount
3122304	end	1101528	mother
3119847	being	1100622	Not
3116084	Next	1100011	memory
3109822	states	1099332	thing
3100333	non	1099257	writing
3099803	even	1099103	none
3077511	project	1098971	screen
3060297	family	1097911	Code
3060145	call	1097416	Journal
3057565	Date	1097249	Day
3057314	Back	1096674	agency

3044343	based	1096532	um
3032906	current	1095750	reading
3030769	al	1095749	November
3029641	pub	1095023	minutes
3022027	van	1092442	france
3021419	science	1091388	items
3016549	Search	1090844	Total
3011683	user	1090529	increase
3010788	place	1090471	tour
3010439	report	1089337	february
3008191	es	1087872	commission
2993346	shall	1087588	schedule
2978125	book	1087549	though
2973138	full	1086904	mode
2968077	John	1086004	forms
2966348	AND	1084873	advanced
2964011	want	1082998	Le
2959642	product	1081998	Council
2942418	does	1081596	account
2933424	down	1080570	practice
2932389	space	1080095	came
2923903	during	1079494	iii
2914326	There	1079080	operating
2912248	real	1077646	TV
2907057	level	1076060	customer
2903721	35	1075910	receive
2897337	small	1074547	FAQ
2891132	while	1073369	nicht
2888535	total	1071885	xbm
2886832	special	1071437	methods
2886578	community	1070753	friends
2881969	link	1069591	pa

2880917	mit	1068832	ups
2877474	Fax	1067393	processing
2875382	Business	1066177	Special
2871957	On	1066105	entry
2871543	HTM	1065994	rules
2867574	american	1064102	higher
2862331	cd	1063394	nature
2859504	much	1062498	ID
2831151	Center	1061936	ce
2820558	board	1060455	facilities
2818649	assets	1059216	Type
2816343	post	1058639	parts
2814622	contents	1058119	editor
2799868	including	1057946	Image
2797855	history	1056700	appropriate
2789452	guide	1055971	calendar
2787693	send	1055696	mary
2784865	college	1055655	Act
2784350	law	1055059	peter
2783565	dot	1053949	Yes
2782824	great	1053813	enough
2780217	code	1052696	centre
2738619	version	1052259	particular
2728188	point	1050588	vol
2727938	Images	1049485	cannot
2723604	government	1048441	average
2718574	files	1046899	gopher
2716225	programs	1046258	certain
2714592	clear	1046193	category
2710565	Company	1045560	February
2701530	include	1045011	til
2694729	water	1044978	modified

2687642	without	1044230	ON
2675773	county	1043706	Pages
2673034	power	1043471	Government
2670885	users	1042591	whole
2666801	own	1039706	option
2665478	read	1039244	Note
2660479	homepage	1038844	sick
2656456	tel	1036288	hall
2640281	another	1034965	effective
2636138	Windows	1034871	prev
2633100	American	1034501	fire
2625906	open	1034018	thought
2623817	provide	1033478	move
2616115	National	1033140	Graphics
2615252	dr	1033112	te
2608924	usa	1032722	distribution
2603000	logo	1032155	sat
2600716	da	1030536	speed
2591558	menu	1029660	input
2587835	But	1029492	sources
2585681	within	1028156	pay
2584602	World	1027269	simple
2581665	directory	1026056	mi
2577737	As	1026042	step
2577330	since	1025888	cover
2570548	zip	1025854	ps
2561971	questions	1025715	academic
2559523	student	1024041	developed
2555581	txt	1023643	lake
2553226	house	1023040	published
2545183	process	1022768	Thomas
2542541	They	1021990	math

C

2540993	pl	1021933	Road
2537582	von	1021055	insurance
2529955	City	1020485	recent
2522966	TO	1020368	growth
2521153	1993	1019913	ha
2519231	table	1018880	makes
2502724	price	1018558	close
2500581	welcome	1017426	physical
2498871	Return	1016908	Rights
2486226	second	1016356	entertainment
2480752	case	1015575	della
2478241	show	1015574	INDEX
2474956	Name	1015417	HOME
2474449	author	1012694	asked
2458453	id	1012366	apple
2454057	value	1012220	maps
2451703	old	1012113	rates
2449726	change	1011430	effect
2448836	box	1011046	master
2440188	staff	1009140	Ltd
2423010	Department	1008552	summer
2419558	ii	1008465	rock
2414136	School	1008193	political
2413828	united	1007383	skills
2410407	did	1006624	Law
2397509	think	1006149	port
2396930	press	1006060	cities
2396141	Research	1005395	needed
2380855	children	1005056	running
2371698	con	1004200	window
2357191	St	1004186	Send
2355166	International	1004184	machine

2354589	When	1004129	Netscape
2350196	come	1003876	hardware
2338114	mr	1003310	band
2330571	red	1003086	numbers
2315604	team	1002885	ads
2312213	est	1002436	brown
2309357	still	1002370	australia
2308252	white	999751	sale
2303454	training	999591	degree
2301258	description	999328	names
2297768	sites	996155	george
2294238	Service	995973	believe
2293231	note	995921	face
2290465	application	994779	gz
2290376	Last	994454	town
2284841	url	994234	turn
2273323	para	993942	rather
2273306	fr	993735	stuff
2273081	person	993089	string
2272041	sports	993005	union
2270824	different	992708	letter
2268245	personal	991836	culture
2265842	times	990444	continue
2262936	States	989725	learn
2261988	messages	989194	His
2261830	Subject	988851	Book
2257310	CD	987416	mind
2256227	every	986842	types
2251330	US	986281	industrial
2250438	How	985146	UK
2247838	Dr	984936	river
2244094	field	984730	cases

2242702	na	983222	contains
2240976	ve	983208	style
2238956	view	983192	lists
2234698	reserved	981810	figure
2233252	das	981531	solutions
2226732	sep	981463	host
2224617	USA	981388	rom
2223314	september	979424	living
2222406	java	978961	oh
2216008	aol	978259	sec
2214211	little	977313	allow
2211554	bar	976781	Conference
2205825	market	976722	soon
2203929	test	976655	basis
2202226	english	975725	Is
2200848	however	975565	fri
2186989	north	975428	agreement
2182779	Mail	974412	NO
2179821	size	974189	lo
2177877	around	974032	Follow
2176388	pics	973907	transfer
2175298	media	973840	door
2172943	example	973539	doing
2165223	pc	973513	programming
2164338	video	972725	Mon
2156120	United	972492	1988
2155613	days	972309	browser
2155187	week	972028	audio
2151521	related	970735	smith
2149233	found	969304	El
2146809	Please	968021	units
2142228	care	966552	Sports

2132261	game	966507	till
2130663	study	965458	Mary
2128529	One	964403	foreign
2125520	start	964073	operation
2124781	Software	963600	jobs
2124626	features	963507	East
2124624	member	963121	frame
2122419	II	963119	europe
2121607	left	962939	campus
2117893	air	962535	kind
2116392	si	960807	Also
2113828	ll	960467	classes
2112523	faq	959241	NY
2110487	required	958935	Arts
2103462	york	958653	heart
2097957	flr	955799	sa
2096678	aug	955649	Area
2095833	june	955326	disk
2094221	today	955177	Line
2078222	County	953202	richard
2077664	September	953184	offers
2076219	action	952689	root
2075621	Education	951679	shows
2074861	HTML	951058	wed
2073643	look	950470	purpose
2073341	cost	949680	capital
2069869	women	948673	getting
2068878	microsoft	948214	station
2068449	add	947308	organizations
2068099	IN	946596	Peter
2063815	review	945360	output
2062979	south	945007	decision

2062729	Top	944243	corporate
2061938	country	944095	File
2061187	etc	942251	OR
2060571	These	941965	treatment
2055318	cs	941841	PM
2054628	results	941415	voice
2049287	General	940520	fast
2043277	plan	939754	connection
2037200	sales	939656	anyone
2034889	exe	939465	dat
2034681	ad	937947	Don
2033936	too	937725	People
2033468	La	937682	White
2031150	quality	936145	fvr
2029654	source	935836	seen
2025866	industry	933197	tue
2023867	System	933055	cross
2023533	shtml	931083	mc
2019906	books	930690	sent
2019745	problem	930078	Control
2012779	canada	929746	vs
2010926	sun	929528	technologies
2010845	pictures	928922	near
2008345	issues	928532	teaching
2007003	ch	928022	db
2001762	button	927932	asp
2001293	Previous	927587	located
2000655	room	927526	va
1999840	1992	927337	profile
1999729	events	927238	florida
1999146	large	927047	vi
1999023	regional	925713	Sie

1998870	er	925353	activity
1998363	Your	925296	auch
1990593	June	924946	beginning
1987090	standard	924301	III
1985025	feedback	923928	hope
1984754	Online	923748	Mark
1984696	Mr	922730	didn
1983913	las	922574	educational
1981196	job	922517	develop
1977895	Systems	922507	spring
1977295	problems	922359	shop
1976674	Other	920415	london
1973304	og	920245	har
1971116	Science	918163	toc
1970952	archive	917935	spacer
1970420	By	916506	NT
1960151	med	916167	Fri
1959837	human	915700	latest
1959014	july	914978	values
1958179	god	914605	contract
1956954	major	914278	double
1955223	Sep	913899	risk
1954415	List	912519	fl
1950368	experience	912150	king
1945936	love	910936	thu
1944411	four	910503	mailing
1942631	College	910241	hour
1941521	applications	910026	mission
1940373	reply	909633	survey
1937772	och	909426	Bill
1937480	hours	909035	bytes
1937383	key	908485	fine

1935353	Computer	907659	anything
1932651	default	907649	island
1929693	im	907097	golf
1928859	An	907002	sn
1926340	Contact	906816	topic
1924679	database	906536	Division
1916947	Phone	906335	role
1914255	performance	905962	Wed
1913215	Click	905787	count
1911359	My	903960	outside
1910568	analysis	903463	notice
1909476	article	902938	AT
1905193	street	902922	thanks
1904527	document	902648	baseball
1903048	nl	901825	won
1902693	engineering	901586	laboratory
1901808	York	901047	career
1896685	common	899843	statement
1894043	ball	899418	Federal
1893203	zu	898403	plans
1891337	ibm	898347	went
1883229	Aug	898045	theory
1881267	Office	897493	statistics
1880050	Tel	897039	cause
1879022	plus	895402	professor
1877569	august	894898	graduate
1876447	san	894768	benefits
1874447	why	894746	dan
1872551	URL	894689	isbn
1872352	association	894640	customers
1871150	model	894538	sciences
1870677	companies	894271	levels

1870058	July	893525	rule
1866715	social	893096	according
1865080	act	892642	across
1863884	check	892471	season
1861667	jan	891970	registration
1853710	important	891875	Product
1852239	west	891844	eng
1848177	say	891554	Access
1841251	let	891545	executive
1838828	going	891178	similar
1838793	technical	890520	Australia
1837293	por	890424	nothing
1825813	blue	890269	lha
1825661	run	890203	ok
1825219	black	889789	George
1822024	men	889283	effects
1820737	working	889074	optional
1817779	language	889050	functions
1813002	provided	887996	Air
1811561	equipment	887722	2000
1811301	series	885967	virginia
1811148	club	885882	forward
1808115	yes	885651	active
1807776	committee	885164	friday
1807340	washington	884653	stop
1805879	av	884497	Chapter
1805714	webmaster	884435	lab
1804290	Group	884106	answer
1801490	david	883863	Manager
1796408	april	883644	prior
1793836	single	881295	exec
1793755	financial	881231	gold

1792198	medical	880732	md
1791308	interest	880726	usr
1789887	pm	880522	buy
1787175	million	879482	protection
1785477	Data	879244	Medical
1783128	created	879116	pre
1782633	give	879044	Form
1781109	travel	878510	potential
1779969	several	877941	selected
1779966	August	877928	Version
1774168	wide	877489	max
1768514	command	877060	fund
1760599	march	875900	Real
1758627	things	875599	TXT
1755777	base	875516	Power
1755441	complete	875498	Java
1754906	few	875428	interested
1754096	president	874929	Central
1750321	better	874854	lead
1747973	God	873815	approach
1747245	Canada	873653	involved
1746617	society	872607	Review
1744076	against	871681	lower
1743783	una	870851	wwwboard
1743670	a	870355	ab
1743476	games	870261	Year
1741398	Program	870226	bad
1740869	again	869446	tree
1740123	possible	868225	IS
1736812	Library	867740	mean
1734311	ou	867607	else
1731088	Links	866460	allows

1729842	Health	864183	nav
1729345	DE	863124	Thu
1728016	Our	862490	sea
1725697	above	861993	channel
1725196	given	861851	almost
1722603	never	860867	COM
1719306	une	859859	started
1717858	issue	859208	corp
1716800	big	859199	computing
1716787	areas	859198	tx
1716059	CA	859110	primary
1714260	live	857490	opportunity
1713820	Management	857060	Each
1710871	activities	856800	Communications
1705327	Site	855609	Media
1704708	studies	855062	save
1703549	het	854387	usually
1701156	future	853860	aus
1700284	location	853596	Studies
1698790	ed	853083	arrow
1698631	reference	852890	1987
1694889	som	851265	Richard
1693520	director	851181	manual
1688280	At	850532	Tue
1687965	request	850426	models
1687241	security	849960	como
1683742	dec	849651	mike
1683419	called	849413	investment
1683417	North	849191	directly
1680146	really	848917	IP
1680141	court	848907	character
1674782	ist	848589	begin

1671762	digital	848432	european
1671750	meeting	847703	took
1671528	provides	847023	rd
1671335	via	845739	scripts
1671053	follow	845720	domain
1667891	federal	844892	Class
1667732	April	844782	Part
1662326	conference	844270	length
1662031	Board	842901	papers
1659304	PC	842320	zur
1656430	WWW	842244	Japan
1655614	building	842233	min
1654683	Guide	842127	physics
1652645	road	842005	Open
1652106	pour	841996	matter
1646799	pe	841791	oder
1646192	body	841191	session
1645884	put	840630	leave
1639720	That	839623	af
1638516	head	838871	multiple
1638405	uni	838608	french
1637464	det	838192	bilder
1637147	policy	837958	register
1636536	sie	837898	package
1634836	less	837658	uses
1631915	object	836603	listed
1631816	talk	836465	western
1631503	She	836170	someone
1630693	1991	836101	blank
1625909	rate	834717	Technical
1624300	FOR	834550	birth
1623310	March	834209	Life

1622036	content	834118	London
1621382	ja	834108	award
1621263	mar	833189	funds
1620610	Jan	833052	Friday
1620525	San	832680	existing
1620357	With	832574	Call
1620323	might	831665	owner
1619386	paper	831374	En
1618911	internal	831329	interactive
1618642	range	831296	associated
1616243	low	830862	thus
1615236	journal	830409	employees
1614402	David	829179	comment
1611107	docs	828979	focus
1609329	until	828725	choice
1608373	corporation	828551	listings
1608064	investor	828011	Two
1606657	thread	827350	FAX
1606146	Help	826951	oil
1604481	below	825947	bei
1602135	About	825818	remote
1601698	california	825519	probably
1597055	updated	825481	mm
1592049	download	823049	completed
1590459	bill	821886	strong
1590131	upon	821845	listing
1589580	jul	821655	pdf
1588829	requirements	821511	Commission
1587485	courses	820752	born
1587484	environment	820531	stories
1585186	word	820487	chicago
1582630	others	819735	surface

1581165	auto	819513	Then
1580142	Technology	819319	Florida
1578546	par	819142	motif
1576067	money	818911	mil
1575577	lot	818700	techniques
1573551	food	817611	cool
1573246	play	817570	cat
1571282	Co	817489	gas
1568635	forum	817212	lost
1564971	Yahoo	817138	medicine
1563065	term	816948	late
1561261	South	816904	therefore
1561228	Network	816537	plant
1561041	att	816016	opportunities
1559376	arts	815747	tu
1556983	additional	814183	beach
1556852	park	814141	movie
1552382	specific	813357	advertising
1551217	English	813185	kan
1550393	needs	812832	solution
1548656	question	812632	bay
1545960	tv	811636	feature
1544716	central	811212	Los
1543251	record	810822	ISBN
1540652	So	810484	official
1540516	includes	810162	track
1537560	electronic	809951	apply
1533541	daily	809406	NEW
1531901	month	808694	hospital
1530778	feb	807276	Text
1530756	always	807221	sample
1530087	catalog	807018	idea

1528100	First	805726	Video
1526198	Association	805595	sign
1525901	ne	804906	dr
1522978	color	804791	adult
1521713	period	804238	reason
1518851	age	804109	monday
1517608	tools	803987	coming
1513331	netscape	803203	ten
1502702	item	803131	3d
1500655	got	801720	Menu
1500056	ms	801712	IT
1498631	Of	800339	ground
1498030	card	800325	century
1497065	property	800027	museum
1496844	projects	799955	vote
1495067	side	799858	testing
1494695	five	798289	Student
1492515	jun	798235	Digital
1491308	star	797662	players
1491105	reports	797309	especially
1490200	Microsoft	797266	everything
1487111	Title	797050	martin
1486342	once	796162	European
1484556	notes	795675	movies
1484473	something	795607	Smith
1483888	hand	795563	Les
1481367	communications	795538	population
1478643	icon	794325	prices
1478357	options	794150	LA
1478351	su	793787	providing
1476172	credit	793291	Ups
1475132	america	793198	doesn

1474069	individual	793004	France
1473478	See	792860	christian
1472707	California	792786	yellow
1471793	Regional	791524	rw
1469932	apr	790752	inside
1469453	Main	790580	networks
1468936	Number	790527	az
1468683	Street	790498	assistance
1467537	present	790205	germany
1467153	Comments	789300	District
1466722	groups	788127	elements
1464881	sound	787929	Europe
1463999	green	787284	sunday
1461643	auf	787265	normal
1461109	Position	787167	cards
1460406	Making	785843	established
1458927	Post	785685	specified
1458564	Time	785667	variety
1458094	Chapter	785503	Reserved
1456667	Light	785475	multimedia
1455564	Changes	785356	player
1454463	Land	785277	bottom
1450450	Final	785057	minimum
1448736	Function	784991	Most
1446316	Production	783553	ability
1445481	Private	783484	procedures
1445383	Hard	782698	parents
1444346	James	782518	foundation
1443662	Further		

Behind the Scenes at AltaVista

Throughout this book, you've read about searching with AltaVista and about the technology that makes AltaVista work. You've learned how AltaVista came about, and have probably tried out some searches on your own. To round out the AltaVista experience, the pictures in this appendix give you a sense of the physical environment that AltaVista's hardware requires.

Note *We were fortunate to tour the AltaVista facilities and see the computers and network connections up close. As a rule, however, AltaVista does not offer public tours because of the need to maintain strict security—to ensure that AltaVista remains as reliable as you've grown accustomed to.*

From the first steps into the building that houses AltaVista, visitors cannot help but notice the security. From the security guard viewing at least 20 monitors, switching among many more cameras, to the heavy doors drawing a crisp line between the front and back (the public and the private) areas of the building. Escorted by Phil Steffora, AltaVista Operations Manager, visitors turn a couple of corners, go through a series of locked doors, continue down a long hall, pass through another secured door, and suddenly see a long room with glass walls and computers packed from wall to wall, as shown here:

Phil casually points out the AltaVista front end (previous photo) that includes the parts of AltaVista that actually receive your query from across the Internet. The following photo of the AltaVista back end shows part of one set of the computers which process your query, find the matches, and return it to you.

Digital's Internet Exchange GIGAswitch

Not long ago, personal computer advocates claimed that the days of mammoth computers sitting in refrigerated rooms, sealed behind multiple levels of security, were over—replaced by personal computers and the Internet, which made computing equally accessible to everyone. Ironically, AltaVista, which in many ways made the Internet accessible to the world, sits in one of those very refrigerated rooms. With the incredible numbers of multi-million dollar servers needed to run the service, nothing else but the refrigeration is a possibility since a key to keeping these machines running is the proper temperature. Each of the separate boxes visible in the racks shown in the previous photo is a component of AltaVista—from query processors, to index servers, to the boxes that support partnerships with other Internet companies.

The array of Digital's GIGAswitch/FDDI routers shown here provide the connections between AltaVista and the Internet. These switches offer standard-setting network throughput and help ensure that AltaVista has connections to the Internet that support the volume of traffic that AltaVista generates.

Digital Internet Exchange Main Hall

A little further through the tour, visitors pass through Digital's Palo Alto Internet Exchange (PAIX), developed by Digital's Network Systems Lab (shown in the following photos). PAIX is a mission-critical, full-time (24 hours a day, 365 days a year) Internet access point for Internet Service Providers (ISPs) and content providers. Some service providers connect through this facility, while others actually locate their servers on these premises for optimal Internet connectivity. Conveniently, AltaVista's servers are one floor away, so these ISPs easily take advantage of Internet connectivity.

Digital's Corporate Gateway

In the PAIX, visitors can see Digital's corporate gateway, which is one of the largest gateways in the world, handling over 45 MBs (Megabytes per second) in traffic to and from the Internet. This gateway supports routing, Digital's security firewall, Usenet news, e-mail, a Web proxy server, many Web servers, the Gatekeeper public FTP archives and, of course, the AltaVista search site.

Where the Cutting Edge Thrives

Simply passing through the AltaVista facilities is an overwhelming experience, even for those familiar with large-scale, high-security, computing operations. From the obvious (omnipresent) security awareness to the concentration of cutting-edge technology, AltaVista's operations mirror the AltaVista service—simply impressive.

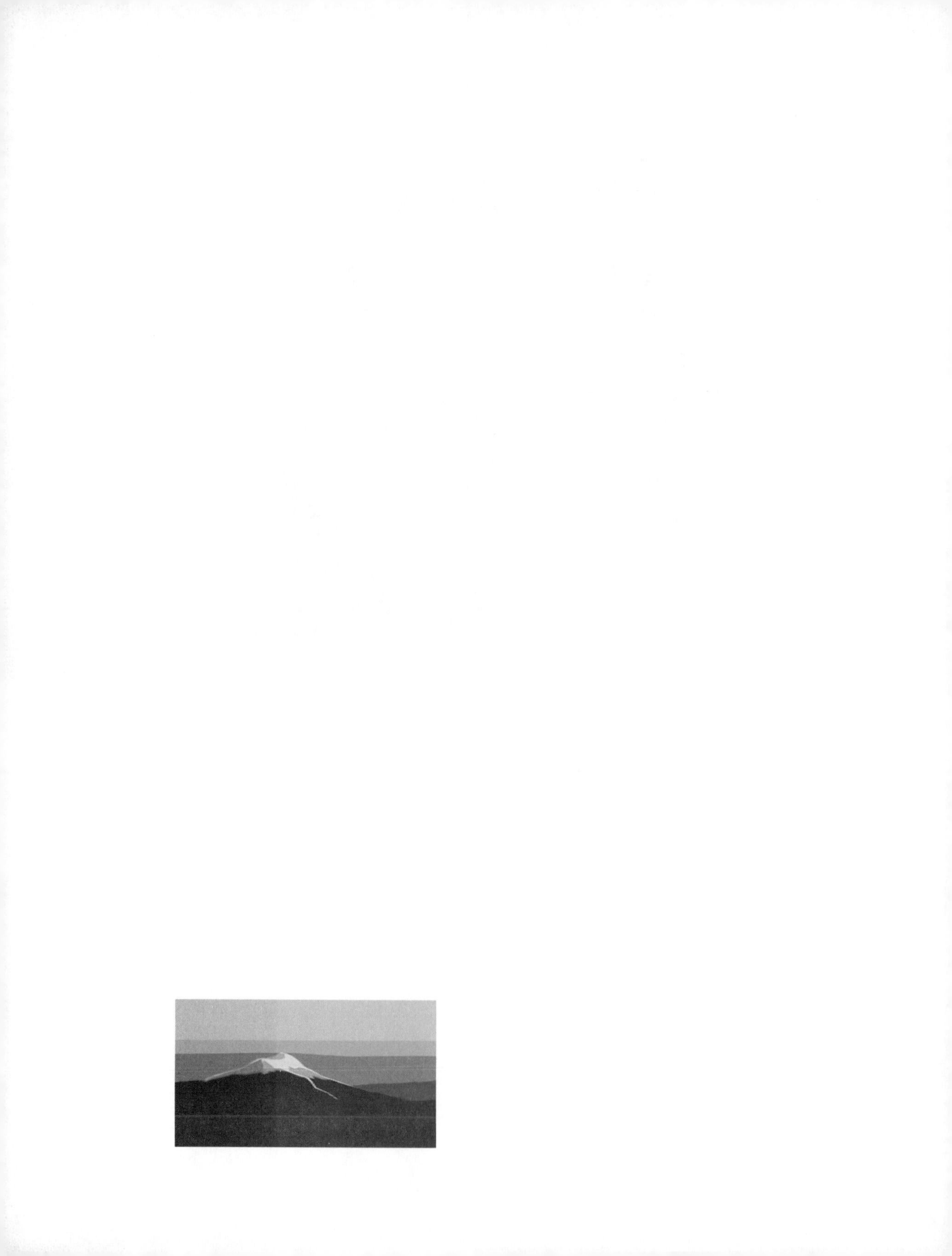

A

B

C

D

E

F

H

I

J

L

M

N

O

P

Q

R

S

AltaVista

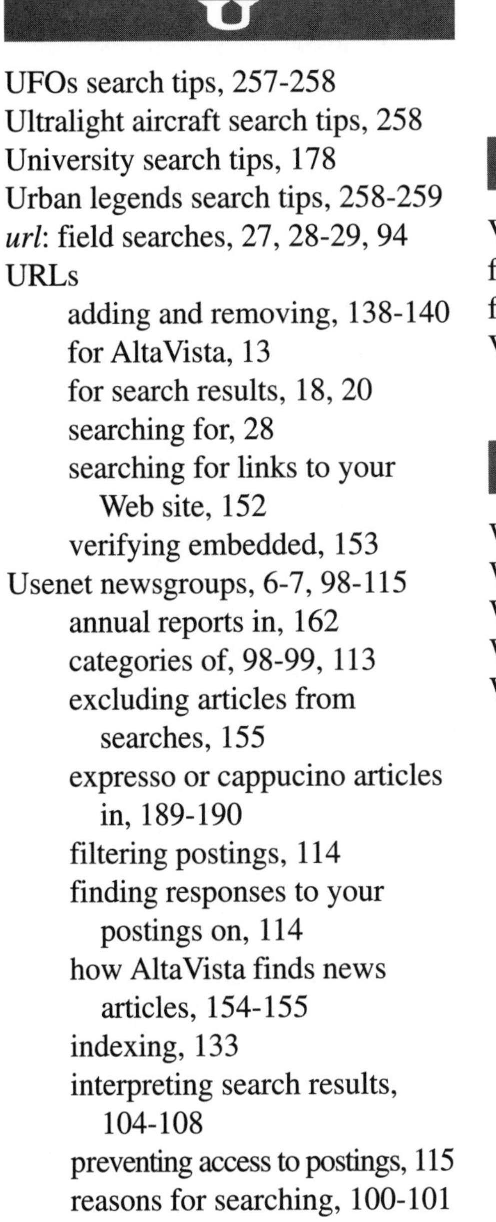

U

V

W

AltaVista

www.cobb.com/isa

INTERNET SEARCH
The Professional's Guide to Internet Searching
advantage

Choosing the right Internet search tool

Search engines provide various methods for accessing information on the Internet. But in order to search the Internet effectively, you must understand and choose the right search tool. In this article, we'll look at the primary Internet search vehicles available and explain the advantages and disadvantages of each type.

Subject directories

Subject directories, like Yahoo!, are the simplest types of Internet databases. They consist of documents that are usually reviewed before being entered into a database and that are identified by category, title, and a few keywords. Because most subject directories manually screen documents, these databases are relatively small compared to large keyword indexes like AltaVista. Subject directories can provide distinct advantages for finding information on the Internet. First, because subject directories are hierarchical in nature, they're great for browsing. Second, the hierarchical nature of subject directories can greatly reduce the odds of irrelevant information showing up in your search results. For example, if you go to Yahoo! and select the Magazines section in the Science category, you know you'll find links to science magazines and not sports or cooking magazines. We've included a list of prominent subject directories in **Table A**.

As the Internet has matured, subject directories have become more sophisticated and therefore even easier to search. For example, Infoseek has implemented special programming controls for Microsoft Internet Explorer, as shown in **Figure A** on the next page. These controls help speed searches by reducing the number of downloads needed to find a precise subject category. Yahoo!, on the other hand, now provides various search options—including

inside

- **AltaVista questions and answers**
- **References and resources**

Table A: *Internet subject directories*

Yahoo!	http://www.yahoo.com
Excite	http://www.excite.com
Infoseek	http://www.infoseek.com
Starting Point	http://www.stpt.com
Magellan	http://mckinley.netcom.com
NetGuide Live	http://www.netguide.com/
WWW Virtual Library	http://vlib.stanford.edu/Overview2

the ability to search only in a specific category—as well as My Yahoo!, a search agent based on technology by Firefly. (See the sections entitled "Personal Spiders" and "Search Agents.")

- **Advantages:** The hierarchical nature of subject directories reduces the number of irrelevant documents. Subject directories are very enjoyable to browse.

Figure A

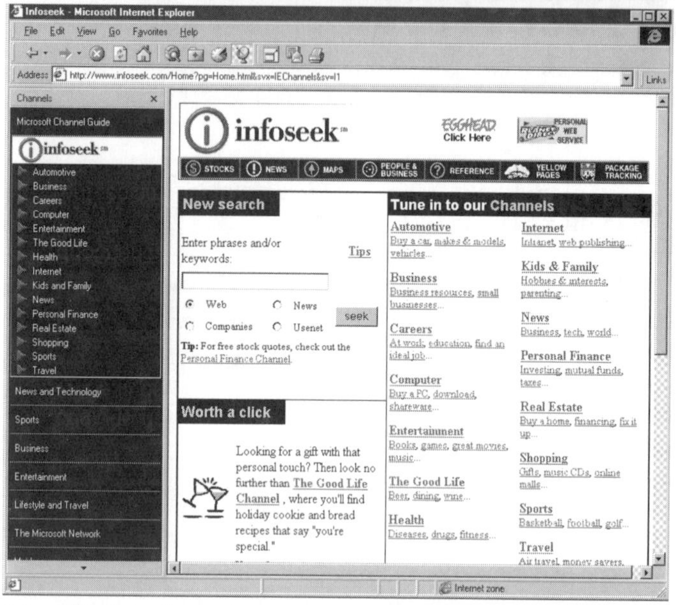

Infoseek's ActiveX page lets you make faster searches while reducing your number of page downloads.

Table B: *Keyword indexes*

AltaVista	http://www.altavista.digital.com
HotBot	http://www.hotbot.com
WebCrawler	http://www.webcrawler.com
Lycos	http://www.lycos.com
Galaxy	http://www.einet.net/
Excite	http://www.excite.com
Northern Light	http://www.nlsearch.com

- **Disadvantage:** Subject directories often contain a relatively small number of Web and Internet documents and tend to be Web-centric.

Keyword indexes

Keyword indexes are large databases that contain literally millions of documents and tens of billions of words. These indexes are so large that robot spiders and indexing computers must build, index, and maintain them. We've listed some of the largest keyword indexes in **Table B**. Let's look at an example of how a keyword index works.

The AltaVista Web site, shown in **Figure B**, is a large keyword index that consists of more than 90 million HTML and USENET documents. The index takes up more than 200 GB of disk space. Maintenance on this index requires the combined capacity of several Digital Alpha servers.

The AltaVista index is constantly rebuilt by a spider named Scooter. *Spiders,* also known as Webcrawlers, are software robots that search for and download documents from the Internet. Scooter can crawl the Web at an amazing rate of three million pages per day. The technology behind Scooter is astounding, but spider technology does have some limitations. For example, Scooter can't retrieve documents behind a server gateway or firewall. In addition, Scooter won't enter servers that request no visits by automated systems. Scooter also has trouble keeping up with timely information, such as stock reports and daily news articles. After Scooter retrieves documents from the Internet, AltaVista's Web Indexer—a computer capable of indexing more than 1 GB of document text per hour—indexes the data.

The real strength of a keyword index is in the combined power of a large database and a fast, effective search engine. Most keyword indexes use pretty standard search procedures for their search engines, but each keyword index has its own nuances and procedures for conducting advanced searches. For example, AltaVista offers an advanced search method that lets you limit searches to specific HTML tags within a Web document. Excite and Web-Crawler offer QBE (Query By Example), which allows you to point out a particularly relevant document in a search result; the search engine then finds similar documents based on a keyword from that document. Learning the advanced search procedures of a keyword index is essential and can mean the difference between searches that result in tens of thousands of hits, as shown earlier in Figure B, and those that result in just a few.

- **Advantage:** Keyword indexes can provide quick access to vast amounts of information.
- **Disadvantages:** To effectively use a keyword index, you must understand its search methods. Keyword indexes have difficulty keeping up with timely information and can't access documents behind server firewalls. Keyword indexes are often Web-centric, providing little or no information about the contents of FTP, Telnet, WAIS, and other specialized databases available via the Internet. The rapid growth of the Internet may soon make a comprehensive index impossible.

Meta search tools

Even the largest keyword indexes don't contain all the information on the Internet, and many tend to specialize in certain types of information. For example, the Galaxy index includes lots of Telnet sites. One good way of making sure your search is complete is by using meta search tools. Meta search tools let you access several databases at once. There are many pseudo meta search tools, such as All 4 One Search Machine, Metasearch, and FindIt!, which are really search link stations that embed multiple search engines on a single page so that you can access them one at a time. These search link stations can prove useful if they're well-designed. For example, the Research-It! Web site at

http://www.iTools.com/research-it/research-it.html

is a great link station for searching dictionary, thesaurus, language translation, and other reference databases. But true meta

Figure B

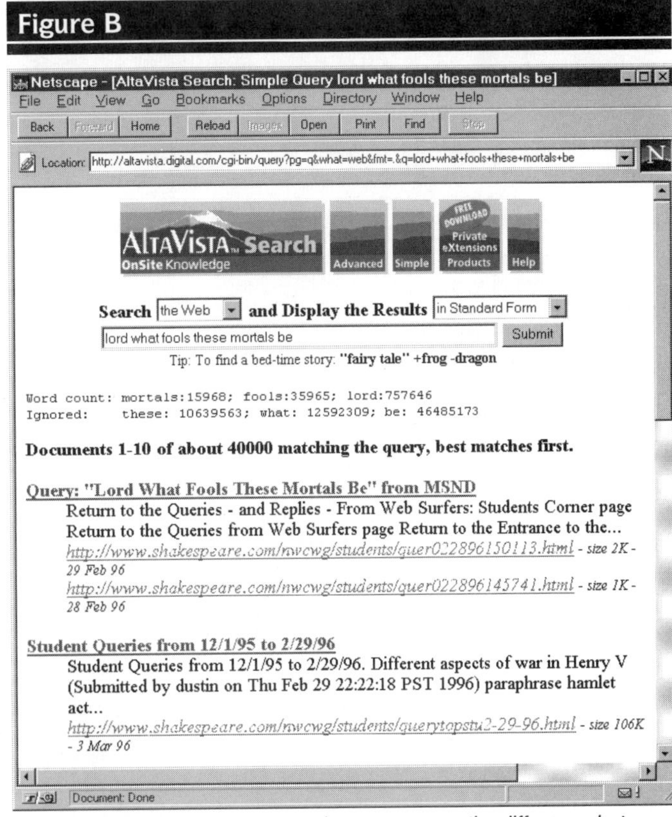

Learning advanced search procedures can mean the difference between tens of thousands of hits or just a few.

Table C: *Meta search engines*

SavvySearch	http://guaraldi.cs.colostate.edu:2000/
The Internet Sleuth	http://www.isleuth.com
Dogpile	http://www.dogpile.com/
MetaCrawler	http://www.metacrawler.com
Highway 61	http://www.highway61.com
SuperSeek	http://w3.superseek.com/superseek/too many frames!
Personal Compass	http://www.personalcompass.com/
Inference Fine	http://www.inference.com/ifind/
Mother Load	http://www.cosmix.com/motherload/insane/
DigiSearch	http://www.digiway.com/digisearch/
ProFusion	http://www.designlab.ukans.edu/profusion/

Table D: *Subject-specific search engines*

BigBook: listings and addresses of more than 16 million businesses in the U.S.	http://www.bigbook.com
JobWeb: A jobs search engine sponsored by the National Association of Colleges and Employers	http://www.jobweb.org/
CollegeNET: an online handbook of information on U.S. colleges and universities	http://www.collegenet.com/ ➥cnmain.html
Switchboard: a directory of residential phone numbers	http://www.switchboard.com/
Image Surfer: Yahoo!'s database for finding online images	http://isurf.yahoo.com
Accommodations Search Engine: A database of more than 13,000 hotels, inns, and bed and breakfasts worldwide	http://accom.finder.co.uk/

search engines, like the ones listed in **Table C**, simultaneously query multiple search engines from one search query and then organize the results by search engine or combine them into one set of returns. Meta search engines can give you more complete search results, but there's a cost. They tend to be slow precisely because they query more than one search engine at a time. Conducting a search with a meta search engine can take a minute or more, as opposed to the 5 or 10 seconds it takes using other search tools. And meta search engines won't let you specify search parameters for individual search engines (most meta search engines allow only simple Boolean operators), so you'll have more results to sift through.

- **Advantage:** Meta search engines conduct more thorough searches because they access multiple keyword indexes.
- **Disadvantages:** Meta search engines tend to respond slower because they access multiple keyword indexes. Since meta search engines access multiple indexes, they provide minimal commands for limiting search results.

Subject-specific search engines

Subject-specific search engines, like the ones shown in **Table D**, search databases that concentrate on one particular topic and often provide better access to information than powerful keyword indexes. For example, let's say you're thinking about buying a Nissan Pathfinder and want to get some information on this sports-utility vehicle before you go to a dealer. If you search for *nissan pathfinder* on AltaVista, you'll get more than a thousand hits that include information on used cars, individual dealers, and photo galleries. But if you access a subject-specific database like the American Auto-Site database **(http://auto-site.com)**, you can save time and start searching for prices and statistics on a specific car right away.

There are literally thousands of subject-specific search engines on the Internet. Three of the best resources for finding subject-specific search engines are The Internet Sleuth, shown in **Figure C**, Search.com, and WebTaxi. These search stations organize hundreds of search engines in easy-to-find subject categories. For example, if you go to Search.com and select Automotive from the Search Subjects column, you'll get a list of database search engines specifically for that subject. Search.com is located at **http://www.search.com/search.html**. WebTaxi is located at **http://www.webtaxi.com**. The Internet Sleuth is located at **http://www.isleuth.com/**.

- **Advantage:** Subject-specific search engines effectively eliminate false hits by concentrating on one subject.
- **Disadvantage:** Subject-specific search engines are sometimes difficult to find.

Figure C

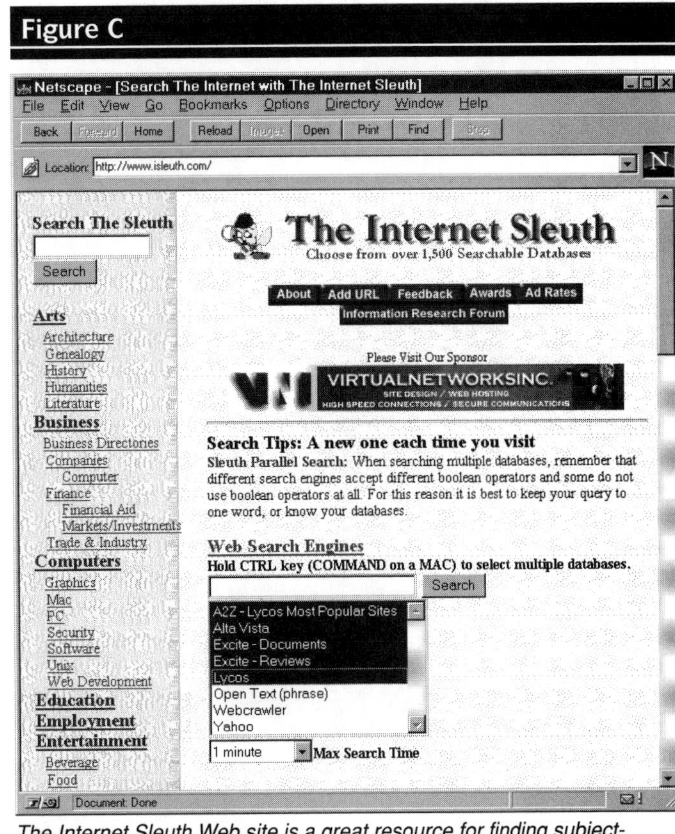

The Internet Sleuth Web site is a great resource for finding subject-specific search engines.

Personal spiders

Personal spiders are a specialized class of meta search utilities that run on your local computer. These programs let you query multiple indexes and directories from your desktop, and you can often configure personal spiders to automatically conduct periodic queries. An early example of a personal spider is URL-minder, which you'll find at

http://www.netmind.com/URL-minder/URL-minder.html

URL-minder is a service that periodically conducts a search for you and then sends you E-mail when new results appear. We've listed some of the more prominent personal spiders and search agents in **Table E**.

- **Advantage:** Personal spiders combine meta searching from your

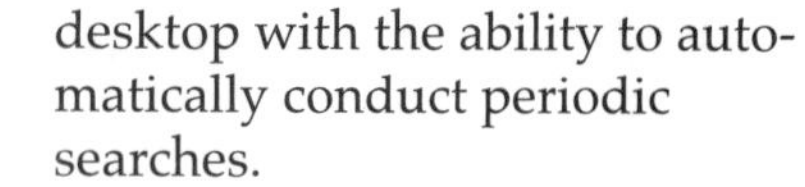

desktop with the ability to automatically conduct periodic searches.

- **Disadvantage:** Personal spiders often cover only one subject and provide only basic Boolean logic for limiting searches.

Table E: *Personal spiders*

EchoSearch	http://www.iconovex.com
Teleport Pro	http://www.tenmax.com/pro.html
WebCompass	http://www.qdeck.com/qdeck/products/wc20/
Surfbot	http://www.surflogic.com/
File Ferret	http://www.ferretsoft.com/
NewsBot	http://www.wired.com/newsbot/
BookWhere? 2000	http://www.bookwhere.com/

Search agents

Search agents are a new class of search engines that use fuzzy logic to let you build on previous searches. (*Fuzzy logic* is a subset of Boolean logic dealing with the concept of partial truths.) Unlike push technology, which lets you configure or select channels that feed categorized news and information, search agents use algorithms that continually refine their understanding of your interests. **Table F** lists three prominent search agents. The Firefly Web site, shown in **Figure D**, is a great example of a search agent that uses fuzzy logic. Firefly combines user profiles and pre-ferences with QBE searching to create communities of interest in music and movies. The more you use Firefly, the better it understands your personal interests. In addition, this agent creates communities of interests that grow. For example, if you like Mozart, Firefly will build on a community that contains more relevant information each time you visit the site.

Figure D

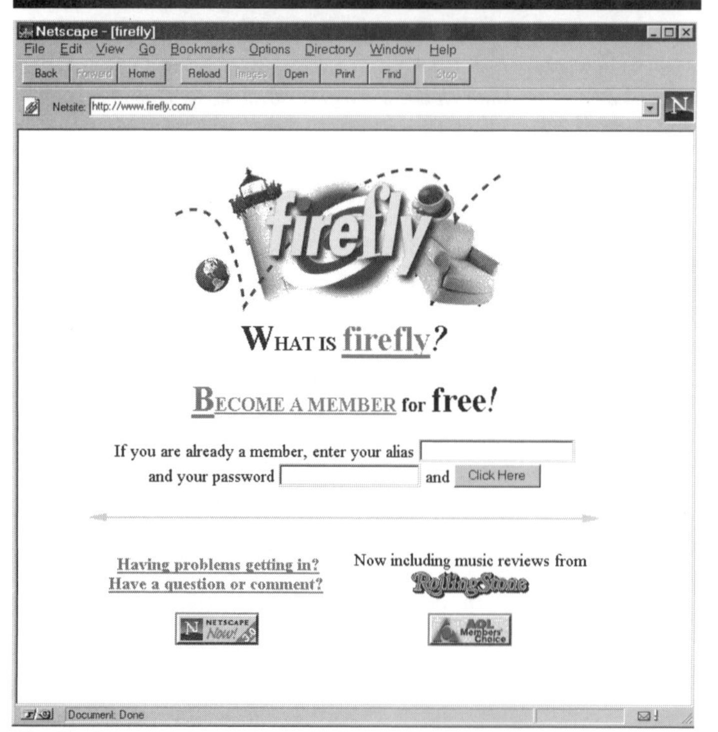

The Firefly search agent uses algorithms that continually refine its understanding of your interests.

- **Advantage:** Search agents help you narrow searches to fit your interests.
- **Disadvantage:** Search agents are relatively new and untested by the general public.

Searching off the Web

The Web is only a subset of the Internet, and although it's growing at a fantastic rate, it still represents only a small percentage of the data available on the Internet as a whole. For example, the major search engines like AltaVista now list about 60 to 70 million Web pages. But the Internet also includes more than 75 million FTP files (see the profile of the search engine Filez); 20,000-plus USENET newsgroups, which account for hundreds of thousands of postings each day; tens of thousands of mailing lists; and thousands of corporate, government, and university databases that are available only via Telnet, WAIS, and Extranet. The search engines listed below are just a sample of Internet resources for searching off the Web for USENET newsgroups, listserv mailing lists, Telnet sites, and FTP files.

Deja News: an archive of USENET news articles
http://www.dejanews.com/

Research.com: A search engine indexing more than 150,000 USENET newsgroups, mailing lists, and Web forums
http://WWW.Reference.COM/

Tile.net: A search engine for listserv mailing lists, USENET newsgroups, FTP sites, computer product vendors, and Internet service and Web design companies
http://www.tile.net/

Table F: *Search agents*

FireFly	http://agentware.com/
Excite	http://www.excite.com/
My Yahoo!	http://my.yahoo.com/

Filez: an index of more than 75 million FTP files on 5,000 FTP servers
http://www.filez.com

Dogpile: a search engine that indexes USENET newsgroups, FTP sites, and newswire sources, as well as the Web
http://www.dogpile.com/

File Ferret: a search agent that searches multiple search indexes and Archie servers for FTP files
http://www.ferretsoft.com

HyTelnet: an index of Telnet sites
http://galaxy.einet.net/hytelnet/HYTELNET.html

- **Advantage:** Searching the Web gives you access to much larger stores of data that aren't otherwise available.
- **Disadvantage:** Searching the Web often takes the extra effort required to understand specialized search engines or search agents.

AltaVista questions and answers

By Richard Seltzer

The following questions are based on recent correspondence about my articles. They deal with matters that many people find confusing, and many of you should find the answers interesting. Please feel free to send your questions to **seltzer@samizdat.com**, or post them in the Forum area in the online version of this publication at **http://www.cobb.com/alt/archive.htm**.

Getting indexed at AltaVista

Q When I edit my Web page, I don't know how to update it in AltaVista, other than by simply uploading with FTP. In my eagerness, I tried going to Add/Remove URL on AltaVista, but I'm afraid I've just messed things up. My page used to come up as number 6 or 7 when I typed *+folsom +homes*. Now, only one of my subsidiary pages appears under the heading TEST. My index page has all but disappeared, as well.

A First, the main criteria for ranking are the HTML title and the first couple of text lines. On each of your pages, make sure you have the most important terms in those two places (the words you expect people to search for). After you've uploaded your edited pages to your Web site, go to AltaVista and use Add URL for each page.

Return to AltaVista after a day or two, search for the terms most important to you (terms you've included in your HTML titles and first lines of text), and see how you rank. You also can search for host: followed by your domain name (if you have your own), or search for url: followed by the directory in which your pages reside (if you don't) to see which of your pages are already included in the index.

A problem with comments

Q I've tried to properly handle the META tags, and for a while, everything was working fine. Then, after looking at document sources on other pages, I decided to add comments like

<!—LISTING="homes/folsom/amriv.jpg"—

I then went to Add URL, and I've had nothing but problems ever since.

A Comments shouldn't make any difference—they're not indexed. As for META tags, if you put what you need in the HTML title and the first couple of text lines, META tags should be unnecessary.

Offending Scooter

Q One thing you suggested really surprised me. You said when I finished editing, I should go to Add URL for each page. I've never done that because when I go to AltaVista's Add URL page, it says, "Please submit only one URL. Our crawler, Scooter, will eventually explore your entire site by following links." I want to be really clear before I start because I certainly don't want to offend Scooter.

A If you want your pages in the index by the next day, use Add URL for each page. If you don't mind waiting several weeks, just enter the home page.

Purging URLs in AltaVista

Q When I was first composing my subsidiary Web pages, I used TEST as the title for pages I was testing. I never added their URLs, and I've since changed the titles. (No, I haven't added their URLs yet, because I thought I wasn't supposed to.) Yet somehow, when I type *+homes +"El Dorado Hills"* I see TEST on the second page and then something about school scores in one of my subsidiary pages or other information I wrote on different pages. For instance, I had a page on climate, a page on recreation, a page on school scores, and so on. But my index page has failed to show up like it once did. So how do I remove all those TEST pages. Do I go to Remove URL?

A Once again, use Add URL and enter the URL of each old, dead TEST page. Scooter will immediately try to fetch the pages. If the pages no longer exist, Scooter will get the message *Error 404* and remove the pages from the index, usually by the next day.

Updating your URLs

Q Now that I've worked with AltaVista as an information provider instead of a researcher, I find that I don't like the search index nearly as much. My job is to list companies in the index, but I've had no success. Is there something I don't know about the AltaVista schedule? Why does AltaVista insist that it updates and has the most current and relevant listings, when in fact it doesn't?

A The information in the AltaVista index is as current as you want to make it. At any time, you can go to the AltaVista site, click Add URL, and simply type the URL of a new or revised page of yours; that info will be in the index usually by the next day.

AltaVista now has more than 90 million pages in its index, and its crawlers visit about 10 million pages a day. So eventually your pages will be found automatically—soon if there are many links from other pages to yours (because that's how the crawler learns about pages—from links), later if there are few links, and perhaps never if there are no links. If being indexed is important, take charge and use Add URL.

Richard Seltzer writes a regular monthly feature on special AltaVista topics for Internet Search Advantage. *He also participates in our message forum. He is a co-author on* The AltaVista Search Revolution, *published by Osborne/McGraw-Hill. His next book is* Internet Advice for Newcomers.

Staff

Editor-in-Chief .... Bruce Spencer
Contributing Editor .... Richard Seltzer
Editors .... Linda Watkins, Laura Merrill
Graphic Designer .... Jeff Dentinger
Managing Editor .... Bob Artner
Circulation Manager .... Dan Scofield
Marketing Director .... Lisa S. Kelley
Group Publisher .... Joelle M. Martin

Prices

Domestic $195/year ($22.50 each) • Outside US $215/year ($24.50 each)
We accept MasterCard, VISA, American Express, and Discover credit cards.

Internet Search Advantage (ISSN pending) is published monthly by The Cobb Group.

Address

You may address tips, special requests, and other correspondence to
The Editor, *Internet Search Advantage*
9420 Bunsen Parkway
Louisville, KY 40220
or send email to isa@cobb.com

For subscriptions, back issues, fulfillment questions, and requests for group subscriptions, address your letters to
Customer Relations
9420 Bunsen Parkway
Louisville, KY 40220
or send email to cobb_customer_relations@zd.com

References and resources

Research-It!

The Research-It! Web site is a one-stop search station of more than 20 reference search engines. This site gives you direct access to dictionaries, thesauruses, acronym databases, quotations, language translators, biographical dictionaries, geographical tools, currency converters, stock quotes, and shipping information. You can find the Research-It! Web site at

http://www.iTools.com/research-it/ ➥research-it.html

Each month, we'll review small search engines, link stations, and reference sites that give excellent resources for your Internet research. If you know of or work for a Web site that you think would be of interest to our readers, please let us know about it. You can contact the editor at **isa@cobb.com**.

Ready Reference

Ready Reference is a link station of reference Web sites covering philosophy, psychology, religion, social studies, languages, natural sciences, mathematics, technology and applied sciences, the arts, literature, geography, and history. The site's subject index is organized using the Dewey Decimal System, and although the site is simple, it's very useful. You'll find the Ready Reference Web site at

http://www.lkwdpl.org/readref.htm

The World's Greatest Speeches

This Web site was developed by AudioNet in conjunction with Softbit to promote a CD-ROM of 400 of the most famous speeches in history. The AudioNet site includes audio versions of some of the most stirring speeches of the twentieth century, including Franklin D. Roosevelt's war message of December 8, 1941; Dr. Martin Luther King, Jr.'s *I Have a Dream* speech; and Edward Kennedy's eulogy for Robert Kennedy. The World's Greatest Speeches site is located at

http://www.audionet.com/speeches/

The Free On-Line Dictionary of Computing (FOLDOC)

This Web site is a dictionary of acronyms, architectures, history, jargon, mathematics, networking, programming, languages, telecommunications information, theory, and tools. It covers just about anything to do with computing and the Internet. FOLDOC is continuously maintained and includes a search engine. You'll find the site at

http://wombat.doc.ic.ac.uk/foldoc/index.html